T0028519

DR ALEXIS WILLETT is a health science writer who spends much of her time turning cutting-edge research and jargon into something meaningful for the public.

also by Alexis Willett

How Much Brain Do We Really Need? (with J. Barnett)
*Drinkology: The Science of What We Drink and
What It Does to Us, from Milks to Martinis*
Science Communication Made Easy (with C. Philpott)
Public Engagement Made Easy (with C. Philpott)
*Rethinking the Brain: Exploring its Capabilities and
How Much We Really Need* (with J. Barnett)

calmism

8 habits for complete rest

DR ALEXIS WILLETT

HEAD START

First published in 2023 by Palazzo Editions Ltd
15 Church Road
London SW13 9HE

www.palazzoeditions.com

Text © 2023 Dr Alexis Willett

Paperback ISBN 9781786750716
eBook ISBN 9781786751485

A CIP catalogue record for this book is available from the British Library.

Design and typesetting by Danny Lyle

Printed in the UK

10 9 8 7 6 5 4 3 2 1

MIX
Paper | Supporting
responsible forestry
FSC
www.fsc.org
FSC® C018072

dedication

To anyone who is feeling overwhelmed, exhausted,
and in desperate need of rest. I hear you.

Also, in memory of Paul Barnes, who sadly passed away during
the writing of this book, and with immense gratitude to Anna
Barnes for unfailingly lending her support despite circumstances.

acknowledgements

I owe enormous thanks to many people who generously contributed their time and wisdom to this project. Firstly, the expert interviewees who answered my inordinate questions with grace and patience: Nicky Dorrington from Sound and Ground, who also kindly provided sound therapy so I could fully experience it beyond my online efforts, Heather Pearson, David Wickett, and Jude Worthington. The Panel volunteers who crammed all sorts of random rest activities into their already busy schedules without complaint (despite not being asked to lie in a hammock or go to a spa, as they had hoped!): Louise, Meg, Rob, Julia, Annalise, Stuart, Caroline, Maria, Graham, Debbie, Katy, Duncan, Jenny, Heather, Marta, Garin, Janine, Liz, Anna, and a very special thank you to Paul who, in the depths of recovery from major surgery and treatment for cancer, enthusiastically took on all the activities he was presented with. You're all incredible people and I'm lucky to know you! Thanks also to Tom Ogden from West Suffolk Council for kindly arranging for me to experience a sound bath at West Stow Anglo-Saxon Village and opening a metaphorical (and literal) gateway to a glorious discovery.

Huge thanks to Duncan Proudfoot for thinking of me for this book and putting up with my requests, and to all the team at Palazzo for driving it to completion. And, of course, it wouldn't have happened without the enduring support of my family and friends, who know that I could do with the advice in this book as much as anyone!

Finally, a quick nod to all those who dedicate their time to helping others find calm and rest – from wellbeing professionals to comedians, musicians to park rangers, teachers to community volunteers, artists to spiritual leaders, writers and many more. Without you, and our precious natural world, the human race would be a hot mess!

"It is not in the stars to hold our destiny, but in ourselves"

William Shakespeare (*Julius Caesar*)

contents

calmism:

building habits for complete rest

"Life seems but a quick succession of busy nothings"
Jane Austen (*Mansfield Park*)

How are you today? Overloaded and exhausted, with a full diary and racing brain? Too busy to think straight? I hear you. Life can sometimes feel like a relentless series of routines from which it seems difficult to escape – work, caring responsibilities, household chores, life administration, social expectations and so on. Even having fun can feel like an effort.

We live in a world that values busyness, efficiency, and success. We're encouraged to do more, learn more, and be more. We're told to set our sights high, aim for the stars, and reach our potential; make the most of every moment and don't hold back. Being constantly on the go and busy, busy, busy is like a badge of honour and, when we have a spare moment, it's now routine to post online to prove how dynamic we are. Everything has become a competition to see who can be busier, be most successful in their career, have the most perfect family, be healthiest, have the most exciting adventures, and own the best things. Living life 'always on, never done' is arduous and unhealthy, but it doesn't have to be this way.

We can, and should, aim to live more restful lives. Turning our attention to what we really need in life, and what helps us feel calm, in control, and fulfilled is the key. Through eight essential habits for complete rest, *calmism* offers a down-to-earth strategy for taking control of your energy and finding that tranquil island inside yourself.

let's talk about exhaustion

Exhaustion (it could be a result of stress, burnout, feeling overwhelmed, anxiety, or simply a very busy lifestyle) can show up in a whole host of ways. (*Warning: long sentence coming up*…) You don't sleep well, you lack mental and physical energy, have little capacity for anything new or non-essential, have low tolerance to stressful events and people, are more emotional, more irritable, experience less enjoyment, feel stuck in a rut, have worse concentration and memory (making you more likely to forget things or make mistakes, subsequently exacerbating your stress), you take longer to recover from mental or physical effort, and/or you may have poor health, from mental ill-health to physical symptoms like digestive problems, loss of appetite, headaches, skin problems, aches and pains, sensitivity to sound, light, and taste, fatigue, dizziness, and breathing and heart problems. Phew! Do you recognise any of these in your own life? There's no denying that being chronically exhausted is bad for us, and it also affects how we live.

A useful way to understand the impact of exhaustion on our lives is described in Mark Williams and Danny Penman's leading book on mindfulness.[1] They highlight the concept of the 'exhaustion funnel'.[2] The widest point, at the top of the funnel, represents a balanced life, filled with the things you need to feel completely sustained. When you become busier, the funnel starts to narrow as you give up elements of your life to focus on what you deem to be important. It becomes increasingly narrow as you become increasingly busy and more stressed. This is because we decide that something's got to give and, unfortunately, we tend to give up more and more of the things that nourish us – like friends, hobbies, and free time. We let go of the things that we see as optional and end up left with work, responsibilities, and other stressors that deplete us of energy, resulting in a relentless downward spiral.

1   *Mindfulness: A Practical Guide to Finding Peace in a Frantic World.*
2   Developed by Marie Åsberg, from Sweden's Karolinska Institute.

When you find yourself in this situation, you may think you just have to plough on or battle through – after all, you're not ill. However, the World Health Organization defines health as 'a state of complete physical, mental and social wellbeing and not merely the absence of disease or infirmity'. In other words, just because you don't have a diagnosed illness, it doesn't mean you're completely healthy. Maybe you just don't feel right or feel yourself – something is missing. It could be that you need complete rest, where every bit of you feels calm, secure, and able to deal with life as it comes.

It can sometimes be difficult to understand why we feel exhausted, especially if we're getting enough sleep or haven't been especially busy lately, and it could be because we're only attending to a few aspects of the rest we need. This was exemplified during the extended lockdowns of the Covid-19 pandemic, which saw a huge wave of exhaustion in otherwise healthy people. Many had fewer demands on their time[3] and more time at home to rest, so what was the problem? Well, we lacked enjoyable events to look forward to, were faced with monotony, couldn't get a change of scene, were restricted in our ability to socialise, became over-sensitised to environmental stimuli, those living alone found themselves isolated whereas those with housemates found it difficult to get quiet time alone, and we floundered as to our purpose in life. In other words, we were missing many aspects of life where we need to find balance to feel truly rested.

what is rest?

Complete rest is somewhat analogous to your general wellbeing. A study from the US described rest as involving a pathway to calm, inner tranquillity, and mental health. Another, from Sweden, suggests that rest is being in harmony in motivation, feeling, and action. Wellbeing is associated with success at personal, professional, and social levels, and there's evidence that people high in wellbeing are more productive, more effective in learning, more creative, have greater social connectedness, and enjoy benefits to physical health and longevity. Prioritising complete rest, and therefore wellbeing, can evidently lead to many additional benefits in life.

---

3   If we ignore the increased caring responsibilities and nightmare that was home-schooling.

Rest is not something we should view as a luxury or a reward after hard work, but as essential for sustaining our health. And we shouldn't sideline it until life forces us to stop. We need to act before reaching the point of burnout. Yes, it may be the wake-up call that some people need but making small changes to daily life when we have the energy and motivation to do so helps offset stresses that will inevitably come further down the line. Life is unpredictable. We won't always know what is waiting for us and whether we'll be able to handle it. When you're completely rested, you feel okay inside and able to cope with life's ups and downs, but when parts of you are in turmoil, you may lack the resilience you really need, and this can fuel a vicious circle. As we become more stressed, we have less tolerance for demanding situations or people, further amplifying our negative responses. For instance, when stressed, we don't sleep so well, we're more likely to make mistakes, forget things, and be short-tempered with others, leading to greater stress and worse sleep, and the more exhausted and overwhelmed we become, the harder it is to think clearly and make rational decisions. When we're depleted, we tend to turn to familiar actions and habits as that may be all we have the energy for, and it can be more difficult to try new things that may help. So, to avoid these issues, we need to rest *before* we're completely worn out. We don't wait for our teeth to rot before we start daily brushing to improve our dental health – we brush our teeth regularly to prevent the rot setting in. In the same way, we need to incorporate positive rest habits in our regular routines to support our long-term wellbeing.

types of rest

Complete rest is like a jigsaw puzzle: pieces that fit together to create a picture of health. If one or more pieces are missing, the picture isn't complete, and the puzzle isn't solved. It's tempting to think that a few good nights of sleep or a vacation will tackle your exhaustion issues but they're just a couple of pieces of the puzzle. Plenty of people get enough sleep but don't feel fully rested, and a holiday to recharge is lovely but you can still find yourself lying on a beautiful beach with a head full of worries. And that's because sleep and holidays are only a part of a wider system that supports our rest needs.

When we think of the need for rest, we usually think of relaxing the body, but we know that feeling exhausted is more than just a physical response. Our minds also need to rest, along with our whole selves, so we need a holistic approach to ensure all angles are covered. A few years ago, Saundra Dalton-Smith, a physician from the US, suggested that seven different types of rest are required to fully restore our energy and maintain balance in our lives: physical rest, mental rest, emotional rest, spiritual rest, social rest, sensory rest, and creative rest. In this book, I offer my own take on these rest areas and propose one more – nutritional rest – and show how we can achieve them by developing positive habits. Delving into the details, I explore what we know about these rest needs, consider a variety of approaches and cultural philosophies that are said to help, and find practical ways in which we can all incorporate regular moments of rest into our daily lives. By developing rest habits in these eight areas, all the pieces of the jigsaw should fall into place, giving you complete rest.

It's important to remember that for any of life's big challenges, the activities in this or any other wellbeing book won't cure the problem – you ultimately have to deal with the root cause – but they may help you to cope a little better, replenish your energy, see things more clearly, and give you the motivation to make changes.

how can *calmism* help?

calmism takes a whole-system approach, incorporating a wide range of ideas and practices that aim to help us feel better and live more contentedly, supported by the science of activities that could benefit us as well as insights from people who've tried them.

Eight habits might sound a lot, but *calmism* is all about finding small, easy ways to introduce more rest into your day, every day, for long-term restoration. You wouldn't eat all your fruits and vegetables on the first day of the month and expect nutritional benefits for the following four weeks, and you wouldn't spend a whole day at the gym and assume you need no further exercise for a long while. We all know that's not the way the body works, and it's no different for wellbeing. A spa break may feel fabulous, but you can't expect it to lead to inner tranquillity for the next few months until you visit again. What you need is to make rest a consistent part of everyday life.

Have you ever noticed that a lot of wellbeing advice assumes that you live alone, with a ton of free time and no outside distractions? This simply doesn't match with real life. Celebrities and influencers often promote lifestyles that are unattainable for most of us. *calmism* takes a more realistic approach, focusing on finding regular moments throughout the week to aid rest – moments that add up to a big difference. With just a few minutes' attention each day, we can cultivate positive daily habits that stick, to help us feel ultimately revitalised.

There's a common adage that every day brings a new beginning – meaning that nothing has to stay the same and tomorrow you have the opportunity to improve. *calmism* takes this further, in that every moment is an opportunity to make a difference to how you feel by consciously being in control of how you choose to think and respond. Although it may not always seem like it, there's usually a choice.

calmism presents a menu of easy-fit options to give you the variety that's essential for wellbeing. It's about finding what works for you and taking your own path to calm.

introducing 'the panel'

For this book, I ran a small (unscientific) experiment to discover whether a range of brief rest activities could make a tangible difference to how people feel and find out how easy they were to incorporate into busy lives. A group of 20 volunteers ('The Panel'), comprising a variety of individuals with diverse lifestyles, living in different locations and settings, tried activities for a week. They included full-time mums, business consultants, teaching, administrative, marketing, publishing, and HR professionals, technology experts, engineers, designers, scientists, doctors, and the retired. They each tried various combinations of activities that were very much entry level – things for a beginner to try and nothing that involved a lot of practice or guidance[4] or that would take up too much time. If

---

4   Such as meditation or yoga, which take time to get the hang of.

people are going to fit in new habits to feel more rested, activities need to be practical in the messy world of real life, so I chose ones that are doable for the average person. You'll notice in the volunteer feedback how they managed to fit things in (or had trouble doing so) around their own circumstances. You might even find you relate to some of their situations.

Over the short time of the experiment, a lot was happening in the volunteers' lives. Along with their regular roles, responsibilities, and stresses, one was recovering from major surgery and treatment for cancer, two caught Covid-19, one had an acute infection, five went on family holidays, and others were juggling big work projects, long commutes, family issues, and changing jobs. Despite the challenges, they all managed to fit some brief activities into their routines – if they could do it, we all can do something!

We knew that a large difference in rest and wellbeing wouldn't be seen in just a week, but any effects would hopefully build over time. You'll find out how they got on throughout the book. They were very honest about their experiences, and gaining benefits was often down to personal needs and preferences. I also tried many of the activities to compare notes and you'll hear about this too. It's worth mentioning that people who know me well would probably say that I'm a bit of a cynic. I'm definitely not someone who you would've thought would be into calming activities, and might even eye-roll at the prospect. But with a racing brain and rubbish sleep, I threw myself in with an open mind and went on a delightful journey of discovery that led me to find calm.

how to use this book

Each rest habit is broken down into possible activities to try, an explanation of the thinking behind them, and views from The Panel. From mindfulness and Zen gardening to cold showers and 'ikigai', yoga and sound bathing to spiritual practices and acts of kindness, there are so many ways to introduce more rest into our lives. This book represents a starting point – think of it like an encyclopaedia of ideas. I'll introduce you to plenty of activities and approaches to try out and you can explore in depth elsewhere those that interest you as well as other complementary tools and techniques. In the final chapter, I suggest tips on how to put positive rest behaviours into practice and transform them into routine habits that fit effortlessly into your daily life.

While some activities may be new to you, some will be unapologetically familiar and seem trivial or entirely un-groundbreaking (is that even a word?!). The point is that they don't have to be revolutionary to be effective. The key is to routinely incorporate these small actions into your lifestyle to feel long-lasting benefits. Just as we look forward to a holiday to take a break from our routines, *calmism* encourages us to take a tiny holiday for the body and soul every day. It's about making an active choice to invest wisely in your time and your self-care.

Now, I know you're tired. After all, isn't that a major reason why you picked this book up in the first place? So, I'm giving you permission to not have to read the whole book. I know that's hard for some of you who feel compelled to finish books, and if you want to read it all, that's brilliant, but don't feel you must. It shouldn't feel like a chore. I don't want you to stress about it or feel overwhelmed by too much information. Read it as you wish. I want you to feel calm and in control of your own choices. That's the point of *calmism*! But, remember, to achieve complete rest, all eight habits need to be integrated into your life, in your own way.

I should also say that I'm not here to tell you what you should do. The information and activities in this book are simply there to help you better understand your rest needs and highlight some potentially helpful options. You can choose any that might be right for you, or even do none of them if you prefer. That's fine too. After all, you're in control. And, while I present the experiences of other people in this book, I know that none of us are you. Just because we found something useful or convenient, it doesn't mean it'll be the same for you. Your life is different to anyone else's and you, and your circumstances, are unique. That's what makes you special! You know you best, so simply go with your gut and try things that feel right. Not everything will be a perfect fit first time, and some might take a bit of practice, so experiment until you find activities that suit your lifestyle and that you can do regularly to transform into positive habits.

Before you begin, just take a moment. Be still. Take a deep breath in. And now let it out slowly and thoughtfully, while your shoulders relax down. Welcome to *calmism*.

habit 1 :

let your brain unwind
for *mental* rest

"It is in your power to withdraw yourself whenever you desire
Perfect tranquillity within consists in the good
ordering of the mind, the realm of your own"
Marcus Aurelius (*Meditations*)

My brain always seems to be on the go, racing from one thought to the next. Whether filled with things I need to do, planning activities, creative ideas, worries, or going over past (or even hypothetical) conversations, my mind is rarely quiet. I know I'm not alone. Are you someone who has a racing or 'chatty' mind that you find difficult to calm? It may come from a need to keep busy or not let anyone down, from perfectionism, too many responsibilities and pressures to juggle, general anxiety, or simply too many ideas firing at once. With the brain constantly under pressure to perform, it can be hard to relax. Even if your brain doesn't race all the time, you may still go through periods when there is just too much to think about or concentrate on at once. Inevitably, this mental overload can take its toll.

When your brain's energy levels are depleted, you feel worn out. Mental exhaustion may show up as extreme tiredness, leaving you feeling drained, irritable, lacking motivation or excitement. It's like your brain is running on empty and can no longer think clearly, so you may be more forgetful, impatient, take longer to retain or process information, have worse concentration, or have poor sleep. You may even end up withdrawing from social engagements or hobbies.

For some of you, this feeling might be a constant state. For others, it may be something that happens now and then, when life gets hectic and overwhelming for a while. If we don't address mental exhaustion, it can lead to more severe health problems such as anxiety, depression, burnout, and physical health disorders. Moving away from the source of the mental overload (e.g. leaving a stressful job) or taking a break can be helpful, but for many people there isn't a single source of stress but a whole heap of things that contribute to a feeling of being overwhelmed. In this case, a break isn't going to make that much of a difference. It all just becomes a bit too much. Too many responsibilities, too many demands on your time, life worries, stresses, and commitments – it all adds up.

It's hard, and not always realistic, to deal with everything at once to make things better, but there are ways to gain more control over the brain's energy levels. We spend a huge amount of time attending to what's on the outside of our head – our skin, our make-up, our hairstyle. On the other hand, we barely give a thought to looking after what's going on inside of it (arguably the far more important bit of our head) until things go wrong. Yet, we can all benefit from helping our brain to be just as well-tended as our external appearance. Developing mental rest habits is critical for our wellbeing.

Before we continue, if you have any mental health concerns, please seek advice from a qualified medical professional for expert support.

You may notice that this chapter is longer than some others and that's because mental rest is so fundamental when it comes to feelings of exhaustion and the need for calm. In addition, it is here that I cover mindfulness, and the science behind it, in detail – a vital approach for helping us gain not only mental rest but other types of rest that we need to find balance in our lives. So, mindfulness will also pop up elsewhere in the book.

what might help?
℮ be mindful

When planning this book, it was clear that I needed to find out what mindfulness is all about. Over the last few years, and especially since the coronavirus pandemic, you can hardly have failed to notice it being promoted everywhere. I knew that it played a role in helping mental wellbeing but how was another matter. So, I signed up for an 8-week mindfulness programme, led by Jude Worthington [see box], and immersed myself in the practice.

Mindfulness, blending ancient meditation practices with modern psychology, encompasses a range of elements around a central theme of paying greater attention to the present; a state of being fully awake to the moment, with acceptance and without judgement. How often does someone ask what you've been up to lately, or what you did in your day, and you simply can't remember? We don't recall events because all the rushing about from one thing to the next means we're going through the motions, on autopilot, without being fully present in our own life. It's slipping by with us hardly noticing. Take a moment to think about when your mind was truly in the present moment. A moment when you were only concerned with what was happening right then. We have a habit of mainly thinking about the past or the future and tend to spend very little time in the present. It's like the child who always wishes they were a little bit older, rushing to grow up rather than enjoying the age they are now.

Being present and embracing ourselves and the world as they are now is the mindful way to fully experience life and, in doing so, reduce fatigue caused by excessive mental energy spent on past and future thoughts. This sounds great, and is definitely something most of us aspire to, but it's easier said than done. Our minds wander from the present all the time – we're thinking about things that took place in the past, replaying conversations and events, planning or looking forward to the future, and worrying about things that may or may not happen. For our brains to get better at staying in the present, it seems we need to practise. We need to train ourselves in how to notice, and engage in, the moment, and mindfulness is an approach that can help us achieve this.

Meditation is a core feature of mindfulness, which I'll come onto later, but let's stick with the concept that we typically go through life moving from one thing to the next, taking things around us for granted. To feel truly rested, we need to better ground ourselves and avoid life running away with itself. Taking time to notice what's happening right now is essential and can be done in lots of ways. Let's take your environment, for example. Look around you – really look. What do you observe? Notice the detail in what you see: whether it's the different colours of leaves or individual veins running through them, or the texture of the sofa you're sitting on. Look at the floor or the sky – what do you notice about them? Then listen for any sounds. Feel for textures and smell the things around you. You get the idea. Mindfulness also encourages you to look at yourself deeply. The idea is to observe and experience things as they are right now, and this process is thought to improve our understanding of ourselves and our needs, encourage a greater appreciation for what we have, improve our quality of life and our wellbeing, and promote a sense of peace. I regularly tried this observance and found birdsong to be louder, foods to taste more vibrant, and, strangely, my body to feel heavier and pulled towards the floor. I didn't expect that being 'more grounded' would feel so literal!

Mindfulness also offers strategies to help you to feel calm, relaxed, and just be. I certainly found this when taking time to stop and focus on my breathing, for instance (more on breathing later), as my shoulders, seemingly almost up at my ears, released tension and softened down. It may not make our problems go away, but a mindful approach can help us to handle them more effectively by shining a light on issues and gaining perspective on how to approach them. In other words, to achieve some control by understanding that we have a choice in how we feel about, and respond to, concerns. As Jude explained:

Mindfulness doesn't get rid of problems but helps to create a mindset to address them with clarity and a healthy perspective. It's like a snow globe – when you shake it, the problem swirls around, difficult to grasp and tackle, but when there's stillness the problem settles down. It hasn't gone away, but the stillness has helped to make the problem more manageable. Mindfulness helps to create that stillness and clarity.

Jude Worthington is a primary school teacher who also teaches mindfulness to children, their parents, and other teaching staff. She describes mindfulness as a tool for developing a sense of awareness, calm, and restfulness. At her school, mindfulness is uniquely embedded throughout the entire culture and curriculum. All staff and children are taught mindfulness, and mindful approaches are employed in every aspect of school life. For the children, it's about helping them tune in to awe and wonder.

We help them to savour the good stuff and joy, as well as help them to manage difficulties. Helping them to know themselves is so important, and building true and honest relationships, without judgement, is key. Mindfulness also brings them a sense of stillness and being in the world. I find that the children in my school are so open and honest about what's going on inside them, and they talk in a highly sophisticated way about their internal world. We also get feedback from secondary schools that our children move on to, that they are perceived as being really grounded and that they know what they're about. That self-awareness is such a gift that will hopefully give them strategies to call upon later in life too.

Jude's own journey into mindfulness started when a psychotherapist working at her school took her on a course about teaching mindfulness to children. She thought this would be a great thing for the kids to learn but didn't realise that it was about to change her own life too. She found that the course opened her eyes to how she could benefit from mindfulness personally and went on to embrace it.

Mindfulness helped me hugely professionally, to bring greater understanding and compassion for the children and colleagues, but also helped me personally. I went from being somebody who had no idea who I was, no feelings or emotions, to now being someone who is aware, open, and non-judgemental. I feel that I truly now have that compassion and empathy that I was lacking in the past, and I can say it has transformed my life. It makes me much more effective as a teacher and a parent.

I used to bury my head in the sand without realising it. People would tell me that I'm so strong – I wasn't; it was just that I didn't let these bad things anywhere near me, as I had put up such barriers. Mindfulness taught me how to engage with issues safely as it gives you the tools to be able to approach difficult things.

Jude suggested simple mindful things you could usefully do if you're just starting out or don't have time to regularly meditate:

Experiencing restfulness from nature has got to be top of the pile. Just stop, and look at something with beginner's eyes, even if it's just a leaf, as though you've never seen it before. Dropping your awareness to your feet, especially when barefoot, can also be very grounding in the moment. Finally, noticing your breath and keeping your attention with it, even for just three or four breaths, can really help still the mind and body.

Much research is invested into understanding mindfulness and its potential uses. It is widely referenced that mindfulness practices can result in physical brain changes, as demonstrated in brain imaging studies particularly focused on mindfulness meditation (more on meditation below), rather than other aspects of mindfulness practice. It seems that mindful meditation can (over time) enhance attention – changes in the anterior cingulate cortex, a brain region associated with attention, are most consistently shown. Effects have also been found in networks linked with emotional regulation and stress reduction, and areas of the brain related to self-awareness have been seen to be altered. Why such changes occur is not yet fully understood, but something is happening.

Mindfulness meditation has been said to help a wide assortment of specific health complaints, but most of the scientific studies into these have been of poor quality and the results unconvincing. There are a few conditions, however, that have been shown to derive significant benefits from mindfulness meditation, including depression, anxiety, and chronic pain. In addition, small studies into mindfulness interventions have demonstrated an array of effects, including general wellbeing, the perception of time slowing down, improved sleep quality, reduction in mental fatigue, improvement in job satisfaction, and reduction in job-related stress. These studies, and many more, are all run in different ways and involve different mindful interventions, as mindfulness is not one thing. This makes it difficult to tease apart what specific aspects of mindfulness (e.g. the mindful perspective on life versus the formal practice of meditation) might be having an effect. While it's hard to pinpoint exactly what is going on, it seems that beneficial effects from a mindful approach are possible and we can all do more to appreciate what's happening now, both within ourselves and around us, to improve our mental rest.

Clearly, it's not practical, nor desirable, to spend all our time 'in the moment' as we would never get anything done or plan appropriately for the future. But we can learn to dwell less on the past and worry less about things to come, and consciously, and routinely, take time to enjoy life as it is. A few mindful minutes each day could do us wonders.

℮ try meditation

What might once have been reserved for particular religions and spiritual practices, meditation is now hugely popular around the world, including among celebrities and sports personalities. Oprah Winfrey, Beyoncé, Paul McCartney, Clint Eastwood, Novak Djokovic, and the late Steve Jobs (CEO of Apple) are just some of the highly successful people who have reported meditating regularly, whether to support their mental wellbeing, boost creativity, or even improve their focus and performance. Meditation is now mainstream and still gaining traction.

Meditation is a broad term comprising many ways to achieve a relaxed state of being. These include mindfulness meditation (in which you widen your conscious awareness), guided meditation (often led by a teacher), mantra meditation (where you quietly repeat a calming word or phrase), qi gong (combining meditation, movement, and breathing exercises), and tai chi (a gentle type of martial arts involving a series of slow movements and deep breathing). The aim of such techniques is to calm the mind, improve wellbeing, and, ultimately, achieve a sense of inner peace.

Meditation practices go back thousands of years and feature in many different cultural traditions. In more recent times, additional forms of meditation, designed to suit modern demands, have also sprung up and there's huge interest in the personal use of meditation techniques to combat stress, feeling overwhelmed, and anxiety, as well as to unearth serenity. The practice of meditation focuses concentration on something in the moment. As you might expect, it typically involves a quiet location, along with controlled breathing, to get you in the right frame of mind to then focus your attention, whether on your body, breathing, thoughts, sounds, or a mantra, for example. According to the Buddhist point of view, meditation is a means of enhancing clarity regarding the patterns and habits of your mind and serenely seeing the true nature of things. The purpose is to take responsibility for your own state of mind and cultivate a more positive way of being. Health advisors often recommend meditation as a self-care approach to improving wellbeing and reducing mental strain.

Scientific investigations into the benefits of meditation suggest that it can help to reduce stress, increase resilience, positive thinking, and empathy, improve sleep, control anxiety, enhance self-awareness, better manage chronic pain, and even improve attention and creativity. However, many of the studies are small or complicated by not accounting for different types of meditation and are not in controlled settings or too preliminary; the evidence is not rigorous and is often reported too optimistically. Having said this, there is a vast amount of anecdotal evidence to show how it works for individuals. Certainly I, and others on my mindfulness course, felt the benefits of regular meditation. Millions of people regularly meditate (an estimated 275 million or more globally) due to the benefits they feel they gain from it, from inner tranquillity to more energy and greater clarity of thought.

So, exactly what is going on in the brain and body to enable us to achieve a calm mind and sense of peace? In short, we still don't have a clear picture and the evidence is patchy, but changes appear largely positive. Various studies report that physical changes in the brain occur in people who regularly meditate; however, the results are not consistent in the types of changes (if any at all) and location of the changes. One finding is shrinkage, or reduced response, of the amygdala. This part of the brain has a number of roles, including as a centre of emotions and memories, particularly around fear, anger, and sadness. It is involved in stress and anxiety and known for regulating our 'fight or flight' response. This decrease in size, or reduced activation, of the amygdala is thought to be linked with a greater ability in meditators to cope with stress. It seems that they are less affected by negative stimuli. In addition, some studies have found regular meditators to have changes in the hippocampus (a part of the brain involved in learning and memory). Meditation has also been shown to lead to activation in brain areas involved in processing information about yourself, self-regulation, focused problem-solving, and something called interoception, which is a sense of how you feel at any given time. Interoception, in which you process signals coming from your internal organs to determine your body's state (e.g. hunger, increased heart rate, anxiety), appears to be linked to our mental wellbeing. There's evidence to suggest that how sensitive we are to interoceptive signals may shape our ability to regulate our emotions, and subsequently how susceptible we may be to stress, anxiety, and depression. It's possible that becoming more attuned to how our body is feeling, through focused meditation, could help us to strengthen our resilience and better cope with everyday stressors. However, to truly benefit, we need to practise often as the effects can take many months to realise.

A fascinating study, led by Richard Davidson, professor of psychology and psychiatry at the University of Wisconsin-Madison, investigated the brain activity of non-meditators versus highly practised meditators (Tibetan Buddhists) to see how they responded to pain from a hot metal plate and the anticipation of the pain. They found that when non-meditators were given the signal to expect a painful stimulus, activity in the brain's pain circuits started firing as if they had received the actual pain; however, in experienced meditators there was no significant change in their brain's pain matrix. When the pain stimulus was then delivered, pain circuits in the brains of both groups responded to the heat, but while this quickly died down to baseline levels in the meditators, the effect persisted in the non-meditators, illustrating secondary brain reactions to a pain that was no longer there (i.e. they were ruminating[5] about it). From this, the researchers concluded that we can learn to control how we respond to external stimuli and that it's a skill that may be nurtured and can enable us to live a happier life. (Note to self: keep practising the meditation!)

Meditation certainly causes something to happen in the brain and body, and this is further evidenced by the less-discussed fact that it can lead to unwanted side effects. Some people experience negative effects, from dizziness, nausea, and hyper-sensitivity to light and sound (something I've experienced) to body pain, visions, anxiety, panic, and some re-experiencing traumatic memories. Mental stillness, in which your mind goes blank and you have few or no thoughts, may also occur, which might be welcome during the period of meditation but outside of that period can be problematic if someone can't think clearly. Even a sense of bliss or euphoria experienced during the meditation, which sounds on the surface like a great effect, can result in a subsequent slump in emotions or agitation and a general feeling of dissatisfaction with routine life. Scientists have even drawn parallels between the effects of certain forms of meditation for highly experienced practitioners and high doses of psychedelic drugs, in that both may produce strong, short-term disruptions in self-consciousness. Observations of adverse effects from meditation are nothing new. According to Willoughby Britton, an expert from Brown University on the effects of meditation, reports of such events go back hundreds of years to Buddhist texts, where different experiences of meditation are noted. Prolonged negative effects from meditation even have a name in Zen traditions: 'Zen sickness' or 'meditation sickness'.

Despite all of this, we know that meditation can be helpful for many people. In fact, based on the limited evidence that has been recorded, the rate of negative impacts associated with meditation practice is similar to

---

5   Ruminating typically refers to continuously thinking about something, or dwelling on negative thoughts on a loop.

that found with other psychological treatments. Also, these risks tend to be increased in people who practise very frequent, intensive periods of meditation (many hours each week) rather than a few minutes a few times a week, which is what most people manage.

If I've put you off, I didn't mean to! The funny thing is I only investigated side effects of meditation because I experienced some myself and wondered what was going on. I do, in fact, think that meditation has a lot to offer, and I found it really helps to calm my racing mind and enables me to drop off to sleep more easily at night. But the important thing to remember is that it doesn't just happen. It really does take some practice to get the hang of, and the best way to start is to try it for a minute or two each day and build up to longer sessions. There are plenty of courses and online resources to guide you. It might be a good idea to start with guided meditations or a structured session of classes to show you how it works and what you need to do. You could try a few different styles at first to see what you're most comfortable with or what fits with your lifestyle. Don't expect results overnight but, if you stick with it, over time you'll find it easier to focus and drop away from your stresses, and you might slowly gain significant improvements in your mental rest.

a note on brain waves

When reading any discussion of meditation, you're bound to come across mention of different brain waves and how these alter depending on your state of awareness. But what does this really mean?

Our brains house around 86 billion neurons – nerve cells that communicate information. This information is passed around all the time in the form of electrical signals. These signals vary in their intensity and frequency, just like waves of information, and can be recorded as brain waves. When scientists measure brain activity, via electroencephalography (EEG), what they're actually looking at are these electrical signals firing between neurons. However, the electrical activity of individual neurons is too small to measure, so it is when many neurons fire similar signals at the same time to form a pattern that currents, or 'waves', can be measured. At any one time, we have a mix of electrical frequencies firing in our brain, associated with different brain-wave states. It's the type that is firing most frequently across the neurons at one time that links with your dominant state. The boundaries are blurred, however, and you pass seamlessly between different states during different times of the day and different situations.

There are five main groups of brain waves, categorised by their frequency and amplitude, associated with varying levels of mental awareness: gamma (high concentration), beta (focused, alert, engaged), alpha (calm, relaxed), theta (drowsy, daydreamy – light sleep or deep relaxation), and delta waves (deep sleep). Low-amplitude, high-frequency waves are correlated with higher alertness levels, whereas high-amplitude, low-frequency waves are linked with more relaxed states. During meditation, we might be in an alpha or theta state and studies have also observed increased delta waves.

There's a lot of interest in trying to manipulate brain waves to promote a sense of calm. For instance, research has explored whether increasing alpha waves, the state in which we experience calm, can help to reduce anxiety and stress. In the 1960s and 70s, there was keen interest in using 'EEG biofeedback' to increase alpha waves in the brain. EEG was used to monitor brain activity while participants undertook activities to induce a calm state. Real-time feedback was given to participants to show what was happening in their brains during the tasks, indicating which were increasing alpha waves. However, interest in this area declined in the 1980s as the technique failed to meet expectations in its usefulness. Many variations of biofeedback techniques have since been developed to help treat a range of neuropsychological disorders, as well as to improve cognitive capabilities, creativity, or relaxation in healthy subjects. The evidence for the effectiveness of biofeedback methods is mixed and larger studies are needed. In the meantime, however, many researchers believe that it may be possible to increase our alpha waves at home, via relaxation practices such as meditation, mindfulness, and deep breathing.

☕ get organised (enough)

Away from mindfulness and meditation, there are other straightforward things we can do to save energy and gain mental rest. The first of these is to get organised. Being organised on a day-to-day basis means we have more room in our heads to think about other things. If we can set in place simple arrangements that help our daily routines, we may feel more in control of what needs doing as well as relieve the mind of constant worries, ideas, or to-dos. For example, having a dedicated place where you put your keys or schoolbags means less time rushing around the house trying to find them when you need to leave, or a regular delivery of groceries scheduled in each week means you don't have to fit in a shopping trip, or tidying little and often avoids letting the mess pile up to a bigger, more stressful job. It's all about putting in place organisational strategies that work for you. Whatever you can do to reduce the number of concerns your brain has to spend energy on will leave a little more energy for more important matters.

There just isn't time to get everything done, is there? With all the things we must do, it can be all too easy to become mentally overloaded. These things whir around in our heads, often when we're trying to relax or get to sleep, increasing our stress and reducing our ability to think clearly and rationally about them. Sometimes, just getting thoughts out of our heads and down onto paper in the form of a to-do list can help. It's a simple way to reduce brain clutter and feel like we're doing something practical. Now, I know what you're thinking. This is really obvious stuff and no one needs to be told it. A lot of people already use to-do lists and still have busy minds. I know I do. But maybe we're not using them in the most effective way that can truly relieve our brains of the mental effort of holding and recalling so much information.

To-do lists allow us to record tasks that need our attention and we can refer to the list at any time and slowly make progress. However, to attain mental rest it's not as simple as dumping thoughts into a list at random. Research has shown that having one long, unstructured list of tasks doesn't help us that much and can, conversely, serve to increase anxiety about how much we need to get done.

For some people, to-do lists can be a reassuring tool that helps them feel organised and in control, bring order to mess, and minimise the risk of forgetting things. But others become a slave to the lists, experiencing anxiety if they haven't manged to tick things off or being rendered powerless by the sheer number of tasks on there. These lists can become a mental effort in themselves: keeping them up to date, feeling like you have to put everything on the list or it won't get done, feeling guilty if you miss something off the list, making them look nice. Creating an organised list, a neat distillation of what needs to be done, can seem like an accomplishment in itself, giving the illusion of productivity even though no actual tasks have yet been completed – a beautifully written list is futile if it doesn't lead to constructive action. To-do lists can be overwhelming to the extent that they, counter-intuitively, discourage us from doing anything at all. We may look at the list and wonder where to begin. Or, just to feel like we've achieved something, we may start adding small, easily doable tasks that don't need to be on there just so we can tick something off when completed. This may be a quick win and short-term mood booster but doesn't help address the growing backlog of tasks. And if you just keep on adding items to the list, you're very unlikely ever to complete it. Ultimately, it can be demoralising. Sometimes lists merely become an instant, visual reminder of our failure to tackle our tasks, which just acts to fuel negative emotions. So we may stop looking at them at all.

We need to get the to-do list in order if we are to reap the most benefit from it. Various studies illustrate that people are more likely to stick to their lists and complete their tasks if they are clear on what needs doing, how, and when, with a clear route of progress. Effective to-do lists are well organised. They prioritise and group tasks, and they break tasks down into achievable components or goals. Instead of listing everything, the first thing is to try limiting your list to just three or four priority items to maintain focus and to get them out of your head. Items should be those things that need doing rather than a wish list. Don't clutter up your list with small tasks that you'll get round to even if they're not on a list (e.g. grocery shopping – you'll soon notice if you haven't got anything to eat!). The next thing is to make sure you set achievable goals – not a list of vague tasks (e.g. do some exercise, or spring clean the house). We know that setting specific targets increases the likelihood of tasks being achieved. So instead of 'do some exercise', you could write 'go for run before breakfast on Tuesday' or 'book in for Friday's Zumba class at gym'. For more complex, lengthier tasks that might be tempting to put off, it can help to break them into more manageable, easier chunks. Instead of 'spring clean house', maybe actions like 'clear out cupboard in bedroom', 'pack away winter clothes', or 'take unwanted books to the charity shop on the high street' would be more effective. The biggest step is just getting started, so making the first few steps quick and easy for yourself gets you on a roll and once you've made some headway it may not seem so challenging. Just as the old adage goes, "How do you eat an elephant? One bite at a time."

Lists can lead to a false sense of progress if only the most pressing, quick, and easy tasks are done, leaving the meatier, more complex ones behind. This means little meaningful progress on some important projects is being achieved. We often become absorbed in the most urgent tasks, so we need to be careful not to let those longer-term tasks build up into problems. Let's say we have certain tasks that need doing this week – arrange a dental appointment, buy a birthday gift – but there are also other tasks (often larger jobs) that need doing at some point but with no specific deadline in mind – fix the garden fence, clear email inbox. The items on the second list will get pushed to the side repeatedly as there will always be more pressing things to do. But this will only store up bigger problems for later as the fence falls down, causing tension with neighbours when your pet dog runs amok, and the inbox becomes even bigger and more unmanageable. So, we have to find ways to organise our to-do lists that take into account both the urgent tasks and others that need to be done eventually. Tasks should preferably be grouped by deadline. Put all the tasks that need completing that day/week together, so you can see them at a glance, and longer-term ones in another group. Then, break down the bigger, longer-term projects into smaller tasks with more immediate deadlines. These smaller objectives can then be added sequentially each time you refresh your list. Instead of listing 'decorate spare room', you could say 'choose paint colour for spare room and identify where to buy it' on one list, followed by 'buy paint, find furniture covers, and book babysitter to occupy kids for weekend of the 22nd' on the next, updated list, and so on. That spare room will be looking gorgeously refreshed before you know it!

Simply writing tasks down also helps us put them into perspective. Seeing them written in front of us can help us more easily estimate how big, small, complex, or time-consuming a task might be and what is needed to complete it. We also notice if we're dragging our heels on certain tasks as they keep rolling over to the next list – why are we avoiding them, and if they haven't been done yet, do they need doing at all? It's essential to delete tasks that are no longer a priority or ones that don't have a clear end point.

So, in summary, don't make your list too long, organise the items by when they need doing and group any that are related in some way (maybe you can kill two birds with one stone, as it were, if you notice several things need doing in the same location or by contacting the same people, for example), create actions that are clear and specific, refresh the list regularly to make sure the tasks it contains are still relevant and those with the greatest priority, and make it work for you – you can always reorganise it to better suit your needs.

Enough about writing to-do lists (who knew there would be so much to say about them?!), and let's think about where to keep them so they're actually used. There are now a huge range of ways to organise your schedule and what needs doing. There are apps and online tools that can arrange your life in aesthetically pleasing ways, with reminders, notifications, and cross-links with other tools. But there can be too many options. In fact, when we use too many tools to get organised (such as keeping task information in a diary, plus using reminders, as well as other organisers and sticky notes), we're more likely to miss key information. This strategy places more pressure on you to know what is where and to check multiple sources to keep on top of everything – more graft for your already overworked brain. Having one central list is better as it's much easier to check and keep updated. But where should this list be?

With people ever more wedded to their mobile phones, you might think they are the best option. However, research suggests that using your phone to organise everything might not be the most effective approach for everyone. Using a notebook is thought to be more successful than technology for focusing our attention without distraction from messages and notifications, for instance, and being able to glance at a paper list is much easier than having to actively find a list on a phone or computer. In addition, the physical space on paper or a notepad can help to visually structure lists, given the confines of how wide or long it might be. The physical act of writing, maybe even colouring, and shaping the list can also help us organise our thoughts as we go along and enable our brains to construct information in a way that we want to see it, as well as help embed the information more effectively. Studies show that you remember a list, or other information, better if you handwrite it rather than type it. (This is a good tip for students or anyone else who is capturing notes or information quickly that they need to remember.) As we make progress and remove or add new tasks to a paper list, we need to physically rewrite the lists, giving us further opportunities to rethink tasks and goals. On a phone, for instance, we tend to leave the list there and may or may not delete tasks. Simply put, it appears that we construct and use paper lists and phone lists in different ways, and that our brains may find it easier to digest and use one paper list rather than several different digital ones used in conjunction with each other. Even if you're not ready to give up your digital tools just yet, streamlining your to-do list within a single location is crucial for reducing cognitive effort and supporting mental rest.

And when it comes to all those jobs you've listed, remember that multitasking is not particularly effective and places greater mental strain on you. Try to methodically tackle your tasks one at a time. Along with mini-breaks in between each task (see below), this approach is better for your focus, more likely to result in the desired outcome, and reduces mental clutter and confusion, helping to better manage mental energy.

℮ note your achievements

Another way that some people recommend to calm the mind is to make a note of what you've already done, rather than what you need to do. The aim is to reflect on your achievements and the positive emotions this may evoke rather than dwell on, and fret about, what's to come. It can be a useful strategy for removing mess from the mind, as well as helping to reduce circling thoughts of past events and conversations, for example, that take up mental energy. However, this isn't an effective strategy for everyone. A small study, published in 2018, compared the effect of five minutes of journaling completed activities versus five minutes of writing a detailed to-do list before bedtime, by healthy young adults. They found that those who made the to-do list fell asleep significantly faster than those who journaled achievements, and the more specifically they wrote their to-do list, the faster they fell asleep. For many of us, bedtime worrying about future tasks that need doing is a big contributor to difficulty dozing off. It seems that getting them down on paper can help to clear the mind and better prepare it for sleep, in comparison with reflection on the day's achievements.

Noting down your mind's preoccupations on paper and leaving them there is an increasingly popular activity for wellbeing. Journaling thoughts and feelings, as well as expressive writing, is widely used as an intervention for people with mental ill-health. Research has found modest benefits of journaling for mental health but the effects are variable, largely because journaling is undertaken in a wide variety of ways and is very personal in its methodology. Let's just say that some people like it, others not so much.

take regular breaks (and don't feel guilty about it)

We should only go so far in this drive for organisation. We need a healthy balance of things to look forward to, as well as free time in the diary to be spontaneous, pause and refresh, or simply enjoy the moment. It's essential for mental rest not to schedule every minute of your life. When we spot some free time in the diary it can be tempting to fill it with another task or activity, in the name of efficiency. Yet, powering through, to clear your task list or fit everyone in, is not a good strategy and may make you less effective overall and impact your mental wellbeing.

In the midst of the Covid-19 pandemic, when video meetings were at their peak, a small study by Microsoft's Human Factors Lab found a striking difference in brain activity when people had back-to-back video meetings versus when they had 10-minute breaks between meetings. EEG data from participants showed that when in non-stop meetings their brains displayed increased beta wave activity, which Microsoft associated with increased stress, and this continued to increase over time. When they were given 10-minute breaks, they were asked to meditate (using the 'Headspace' app) rather than fill their time with alternative work tasks. These breaks resulted in a drop in beta wave activity, which was interpreted as the brain having a chance to 'reset', and allowed participants to enter the next meeting fresh. In addition, they noticed brain patterns that indicated higher engagement when participants had the breaks, and less engagement in the meetings without breaks. From this study, the researchers concluded three things: 1) breaks between meetings allow the brain to reset and reduce the cumulative build-up of stress across meetings; 2) non-stop meetings can decrease the ability to focus and engage; and 3) transitioning between meetings can be a source of high stress (there was a peak of beta wave activity during transitions). While this was only a tiny study, it may reveal something about the pressure on the brain when we're always on the go. Other research has also shown that incorporating rest breaks into mentally demanding tasks not only allows for relaxation but also reinvigorates energy levels and decreases fatigue, enhancing performance. In short, it is not healthy or effective to fill every minute of our time with tasks just because we practically, physically, or logistically can.

We often perceive advances in technology to have freed up more of our time as they automate tasks that previously would have been time-consuming. However, as Craig Lambert (author of *Shadow Work: The Unpaid, Unseen Jobs that Fill Your Day*) observes, the opposite is often the case. Just think of those tasks that regularly eat into your time that you would not have done in the past. From scanning your own groceries at the supermarket, online airport check-ins, making event bookings online, registering for online accounts and completing digital forms, or filling up your car at the petrol station. While this is all portrayed as being for your convenience, businesses are readily taking up your free time to benefit their profits rather than investing in more staff or resources. Before you know it, hours of your time have been spent on previously unnecessary tasks – a burgeoning pile of life administration. In addition, we're enabled to do things 24/7, keeping the pressure on us to get things done. Bit by bit our free time is encroached upon, leaving us less and less time for doing things that support our wellbeing and sense of rest.

So, we need to set boundaries on our time. Build regular breaks and protected time into your daily routines, and don't feel guilty about disconnecting from technology (there's no rule that we have to be available at all times for everyone!). Remind yourself that things in life are no more urgent than they were before email and mobile phones were invented. Just because it's possible to do things more quickly (e.g. respond to messages and invitations or send reports to work colleagues at pace) it shouldn't mean we have to. I'm sorry to say that a lot of what most of us spend our time doing isn't lifesaving, groundbreaking, or even essential to the human race or natural world, so why is there so much pressure to do things urgently? There is very little rational justification for this behaviour. It merely serves to trap us in a never-ending cycle of stress, which doesn't do us any good and also doesn't improve performance. If anything, it makes us worse at what we do.

When you find yourself with some free time, try not to immediately fill it with tasks – there will always be things that need doing. Ask yourself, is it critical, urgent, or an emergency? If not, it can probably wait – even if only for 15 minutes.

◉ slow down and look forward to enjoyable events

As well as taking breaks, we can learn to take life at a slower pace at times. By incorporating moments in the day to slow things down, whether it's taking time to sit still for a few minutes, making a cup of tea or a meal slowly, or slowing your pace when walking somewhere, you can help to reduce the frantic rate of your thoughts and think more clearly. Being busy can be energising to a point, where we run on adrenaline to move swiftly, but there's a limit to what we can effectively achieve like this. We need to allow ourselves, and our brains, a change of pace; to come down and settle, allowing our thoughts to align in some sort of order, just like Jude's snow globe analogy. Mindfulness is particularly helpful for this but, even without trying to be 'in the moment', deliberately stopping or slowing down for a few moments helps us catch our breath and refocus.

It may not be desirable to schedule our whole lives, but it's a good idea to get a few enjoyable things in the diary to look forward to, which often requires some planning. Several studies highlight that the anticipation of an enjoyable event can boost positive feelings, sometimes more than the event itself. Holidays are a classic example. Researchers have found that people who take holidays report increased happiness in advance of the holiday, but after their trip don't appear to be significantly happier than those who didn't have a holiday. It seems that beneficial mood effects of vacationing come largely from the anticipation of it, and any benefits from the holidays themselves only appear to last for a week or two afterwards. This is something that got people down during the coronavirus pandemic – there was very little that they could plan or look forward to at the time.

ⓔ acknowledge that you can only do so much

Do you find yourself taking on too much? Are you anxious to do everything just right? Do you worry about letting people down? You may be a people-pleaser, high achiever, and/or something of a perfectionist, and you'll find yourself in great company. Many of the world's most successful people would relate to this description. But it can be wearying to live life like this. You may achieve outstanding results (brilliant work outputs, parent of the year accolade, amazing social life, enviable image etc) but at what cost? Maintaining such high standards places enormous pressure on you to deliver but, while you worry about what other people think, often the person who cares most about it is you. Humans have this terrible habit of creating unnecessary stress for themselves. Have you ever stopped to consider what would happen, or who would notice, if something you did wasn't quite as perfect as usual or didn't happen at all? We might not like to say no, or worry that we're letting someone down, but we only have the capacity to do so much without consequences for our mental rest. It's essential to recognise our own limits and it's up to us to try to manage these. A conversation explaining your situation or suggesting an alternative, less-demanding approach is often all it takes to lessen the burden and resolve the anxiety. You may feel mortified if you don't deliver a high standard in everything you do but, unless it is life and death stuff, does it really matter? It's important to remind yourself that most people don't, and can't, do everything perfectly every day and yet the world keeps on turning. I'll discuss this topic more in habit 2, but in the meantime, allow yourself to think 'I will do my best but can't do everything, and that's enough'. And shed the guilt. Practise this often enough and it will eventually start to feel natural.

Worrying about the unknown, including opinions, is a common phenomenon. Sometimes we do this to feel prepared to handle situations that may arise by imagining them in advance. But this can also lead to catastrophising – where we blow things out of proportion and fuel our own anxiety about things that may never happen. It's a lot of (often pointless) mental effort. When you find yourself in this situation it can help to either arm yourself with the relevant information to rationalise your response to it (what do you really know about the issue or situation?) or simply acknowledge that some things are out of your control and don't concern yourself with them. Understanding and acknowledging what we can control and what we can't is a crucial skill for mental rest. Moving the issue forward by planning, making decisions, or taking action about those things in our control (rather than starting the cycle of worry again) can help increase our readiness and resilience for when things happen. For everything else that's out of our control, we should let it go. (If you now have *that* song from *Frozen* in your head, my apologies.)

what did *the panel* try?

1. Note down preoccupations each day, set them aside, then do something enjoyable immediately afterwards. On the whole, The Panel found this to be somewhat onerous, but some benefits were felt.

At first, I was getting stressed about when would I do it, how tired I felt to do it at the end of the day, and had I got time. But I actually enjoyed it. It was so tempting to turn my preoccupations into a to-do list, because my preoccupations are generally around what I haven't got done that day. So, I turned it into a free-writing exercise. Just a few minutes writing down the kind of thoughts/ judgements around my need to get everything done and how it feels in the body, for example, in the hope that acknowledging this would allow me to let it go, really helped. My five minutes of doing something nice after the writing was one of my fave evening meditations, around self-love. My phrases in my meditation are: 'I've done my best, forget the rest' and 'I am enough and I do enough', so that built on letting go of my to-do list in my writing. After doing this writing/meditation combination, I'm sure I went to bed with a less busy mind. On another night, after writing, I did stretching instead. I probably won't keep this up every night long term but I can see that it is helpful for me and now I have a tool to use when I need to. I really do think it can help with calming the busy mind, getting things into perspective, and seeing how hard I'm being on myself – **Louise.**

I prefer to do this early in the morning while walking our dog. Doing this, while also practising controlled breathing, helped to set the day in a calmer mindset. If I try this at the end of my day, it tends to affect my sleep – **Marta.**

Making a note of my mind's preoccupations just made me want to put a full to-do list together. I think because I am such a list/spreadsheet person, this is how I like to organise, and I just couldn't write things down without having the control of a list. Once in a list, I can then remove them from my worries – **Liz.**

2. Journal your achievements, no matter how big or small, daily, and read them back to yourself. When achievements were captured simply as a reflective moment to learn from, either about the day or themselves, they had a positive impact.

I really loved this! I am a huge list-maker (with four young kids, I have to be!) and I get a sense of achievement from crossing items off my to-do list. Whilst this is similar in a way, I found it to be a much better sense of achievement because it wasn't things I had needed to accomplish, just small things I did throughout the day that I was proud of. I think it's lovely because while I might receive praise or thanks from others for accomplishing big things, these are smaller things that others would not thank me for doing, so it's nice to take a moment and praise myself for them! At the beginning of the week, I started a draft email on my phone and every day I recorded three or four things. This made it easy to write the points down and it was also nice to be able to not just re-read my points from the day, but also each day throughout the week. This was such a pleasant thing to do that I didn't find it inconvenient or hard to get around to doing. I would definitely recommend this activity to others and will continue to do it – **Meg.**

I felt slightly hassled that there was another thing that I needed to think about: 'Is such-and-such an achievement? Shall I put that down? What is an achievement? Have I achieved anything today...?!', and then make a note of it somewhere and not lose it amongst everything else! And then, reflecting on the list that I did make, it struck me that everything on it was stuff that I had done mainly for other people's benefit rather than anything that I would consider as an achievement for myself – **Stuart.**

3. Take a 15-minute complete break during the day, with no distractions (no phones, screens, chores etc). This one was hard for ruminators and those who like to be rushing about. Even if they tried to have a break, their minds would continue to race. But these are the people who need this type of rest the most, to reduce mental exhaustion. A little perseverance and commitment to keep trying started to reap rewards.

I ate lunch on my own in silence, which was okay, but I'm not sure doing it in a work environment was a good idea as I kept thinking about all the things I still needed to do and so wasn't feeling restful, although it did get me to be more mindful about what I was eating. I got to have lunch a few times in my garden the following week (working from home). It enabled me to feel ready for the next half of the workday. I do like to have a quiet cup of coffee in my garden early in the morning though. That is one of my favourite things to do and is both restful and soothing – **Anna.**

I decided to combine my break with lighting a candle and setting a timer to prevent myself getting distracted by work-related stuff. So I lit my candle, set a timer, and sat with my book. The number of times I went to pick up my phone just to do an invoice, look up something, or check my to-do list was huge but the timer brought me back to my book. My neighbour's cat even popped in to say hello and actually jumped up to sit on my lap for a bit. I'm convinced the convivial atmosphere from the candle and my restful ease encouraged him to relax! We sat together and I really enjoyed it. The next day I found this easier. I had lunch and then picked up my book. I didn't time myself, but I felt that however long it was, was the rest that I needed – **Louise.**

This was tricky as I had Covid but continued to work from home. The additional push of having a real break at lunchtime (which I very rarely do) was a good thing. I would stop everything, have lunch, and then take a walk around the garden. I definitely think it helped to ensure I made it through the afternoon without feeling too exhausted – **Liz.**

I find it more relaxing to watch TV when having lunch (if working from home), otherwise I continue to think about work. I tried it at the weekend when not concentrating on work and it worked well then – **Marta.**

4. Meditate. While no Panel member was specifically asked to try this, Louise has been practising this for some time (outside of the experiment). She finds it beneficial for her wellbeing.

Now I've been practising meditation for a while, it's like my mind and body know that it's time for some self-care and peace of mind. I sit and as I concentrate on my breathing, my mind settles and my body regulates. It really has encouraged me to carve out time in my day for just being, whereas before I could never stop doing. I can get into a habit of working without taking a break and ruminating constantly, being unaware of the small details of life. So, when I stop, I'm putting myself in the present moment, calming everything down, and gaining perspective – **Louise.**

You'll see the results of other mindful activities elsewhere in the book.

habit 2:

process your feelings
for *emotional* rest

"Unexpressed emotions will never die
They are buried alive and will come forth later in uglier ways"
Sigmund Freud

You may have come across the Danish concept of *lykke*. It means happiness (something we all strive for in some way). The author of the best-selling *The Little Book of Hygge*, Meik Wiking, went on a search for what makes nations around the world happy and found six common elements. These included togetherness (spending time with other people), prioritising health, having a good degree of freedom (having control over how we spend our time), and showing kindness to others. Societies that scored highly on the six pillars also scored highly on measures of happiness, suggesting that we can learn to be happier by incorporating these elements in our lives. These themes tie in with different aspects of rest and wellbeing, so you'll come across them throughout *calmism*. Yet, while the pursuit of happiness is a noble ambition, emotional rest comes from understanding, processing, and accepting the whole range of our feelings.

Our emotions and feelings can be many things – joy, sadness, frustration, excitement, contentment, fear, surprise, disgust, anger etc. They may be written all over our faces, but can also be mixed, complex, and difficult to understand or express (grief, love, jealousy, guilt). Emotions can be the result of how we think about something (if we remember a loved one who has passed away we might feel sad, or anticipating a pleasurable event might make us feel excited) but can also be the cause of other sensations (being angry might make our stomach hurt, give us a headache, or even make us feel guilty or sad afterwards). They can come out of the blue (a negative thought might pop up or something might happen that makes us feel afraid), or they might appear in a predictable pattern (we might spend time with certain people because they make us feel joyful and loved, or we know that others will bring our mood down).

Emotions have their pros and cons, but they are also a part of what make us human. Important for communicating and bonding with others, allowing us to build relationships. Helping us avoid danger, or things we don't like, by recognising fear or disgust, for instance, and preparing us to take action. This is very useful if you're in a combat situation or under attack, for example. Emotions can help us make decisions based on how we feel, as well as motivate us to act. We all know, however, that emotion is more than a mental phenomenon. It affects the whole of our physical and mental self and way of being, and can overwhelm us in the moment, over time, or even when we're least expecting it. Understandably, how we feel has a huge impact on our lives and getting a handle on our emotions can help us better live life on our own terms and find rest.

what might help?
❻ recognise and process your emotions

I'm guessing most of you have at some point ticked a box stating: 'I am not a robot'. And you would be correct. We're not robots. We all have feelings and emotions[6] and being able to express and regulate them appropriately is essential to our wellbeing. There are said to be three basic elements to our emotions, and each element can help us better understand what we're feeling and why. First, the subjective experience – how we experience the emotion (each of us experiences life, and things that provoke emotions, differently);

6   Emotions occur first as a reaction to a trigger, then feelings result as a conscious, subjective appraisal of what we're experiencing.

second, the physiological response to the emotion – our nervous system's reaction (I'll talk more about the role of the nervous system shortly); and finally, the behavioural response – how we express the emotion and behave in response to it. Simply put, something triggers an emotion and we feel it, setting off a series of physiological and behavioural responses to it.

Broadly speaking, the purpose of emotions is to convey a message to our consciousness, enabling the brain and body to act as necessary. For instance, if we experience positive emotions, we may want to do more of the thing that made us happy. If we experience negative emotions, we may take action to try to avoid whatever caused the unhappiness or minimise its harms. So, we first need to tune in to our emotions to understand what we are feeling.

We experience many emotions and feelings, positive, negative, and neutral, every day. We often start the day in one frame of mind and end in another. Trying to identify why we're feeling an emotion is key to understanding our response to the world, which, in turn, is important for recognising how best to deal with it. We can all think of times when we might have overreacted to a situation. Perhaps we shouted at someone for something really minor, burst into tears at something seemingly innocuous, or laughed incongruously at something not that funny. When we're calm and rational, we may look back on these instances and realise that there was something else going on in our heads at the time that wasn't necessarily related to our outburst. Perhaps we were upset, angry, or anxious about another matter entirely. If that's the case, we can begin to address the trigger that caused these emotions in the first place. That's our starting point to help us take control of our emotional rest.

It's often said that we adopt many of our behaviours in dealing with emotions as children, which might include hiding feelings (in case people laugh at us), lashing out in a temper (when suddenly enraged by something), or ignoring or burying feelings (as a shield to protect ourselves). But the problem for many of us is that we don't then change these behaviours as we mature. Every other part of us grows up but we don't always recognise that we need to advance our approach to emotions too. Hiding feelings or ignoring them doesn't get us very far as adults – while it can protect us for a while, we're essentially just kicking the problem down the road and building up physical and mental harms as a result. Just imagine feeling increasingly frustrated with someone but never expressing it for fear of their, or others', reactions. This is a common occurrence as most of us are more comfortable with keeping the peace and trying to be pleasant. But the frustration can accumulate until one day we explode in a fit of anger, whether at that person or an innocent party – perhaps we end up shouting at a partner or child when we get back from a stressful day interacting with the person who winds us up. It all becomes too much. In hindsight, it would have been better to try and resolve the issue at a much earlier stage. By leaving the issue to fester we could end up with a bigger problem, or even a crisis, on our hands.

It's long been thought that burying, or failing to express, our true feelings is bad for our health and there is some science to back this up. For example, a US study following members of the public over 12 years found that emotional suppression was associated with an increased risk of early death. Other researchers also found links between repressing feelings and negative health impacts. Ignoring our emotions can lead to stress on the whole body, such as raised blood pressure, frustration, memory and self-esteem issues, and potentially anger, anxiety, or depression. Conversely, emotional intelligence, or being in tune with feelings, has been found to be positively associated with psychological wellbeing.

Various practices aim to help us handle our feelings by allowing our emotions to be what they are, processing why we are feeling them, and taking better control over managing them or living with them. Mindfulness is one such example that has been shown to improve emotional regulation and acceptance, and research indicates that people who meditate have greater emotional acceptance than non-meditators. In his book, *Permission to Feel*, author Marc Brackett suggests the RULER approach to handling emotions. Start by **R**ecognising and **U**nderstanding your emotions by using all your senses and cues around you to work out what's going on. Then **L**abel strong emotions with specific words to help break them down (e.g. furious, irritated, scared, worried), as articulating emotions can help you make sense of them and find appropriate help. **E**xpress your feelings to show what's going on underneath the bad mood. And finally, **R**egulate your emotions by being able to anticipate what triggers strong emotions in you and create strategies for dealing with them when they arise.

Cognitive reframing, a technique used to alter mindsets, can help change the way in which we see the world by looking at things from a different angle. This reframing can help liberate us from anxiety, fear, anger, or resentment, for instance. For example, you may have a fear that you will fail your driving test so you keep putting it off. Instead of going round in a negative loop, you might instead ask yourself why you think that – what are the facts? Are you genuinely listening to the instructor who may have faith in you or doubting your own abilities without just cause? This is a fairly minor example, but the technique is also used by professional therapists to address major life issues. Instead of trying to escape something negative, cognitive reframing might focus on it to turn it into something positive or at least offer an opportunity to see it in a less harmful light. This approach can change how we think or feel about people and situations.

Expressive writing, where you write about traumatic, stressful, or emotional experiences, is another technique that has been found to help mental wellbeing, although how it is doing so is unclear. Perhaps simply facing the emotions helps the individual to move them forward, rather than bury them. It's possible that writing something down on paper and transferring it out of the mind helps a person to process the feelings associated with the subject. Perhaps it is giving them space to reflect on what happened and how they're feeling as they write, which may give them further insight and perspective and the opportunity to restructure their thoughts. Perhaps it helps to clear the mind, in part, as it no longer has to hold all the information about the experience.

Some people find the 'Emotional Freedom Technique' helpful in processing their emotions. This technique involves stating the problem that is on your mind, then stimulating acupressure points without needles (using fingers to apply pressure or tapping) in a particular sequence while repeating positive phrases. This is often done under the guidance of a trained therapist but may also be tried on yourself once you've learnt the technique. While it is still a relatively novel therapy, there's some evidence to suggest that it helps relieve anxiety and result in a more positive frame of mind.

It's important to note that if you ever feel taken over by your emotions and find it difficult to understand them or manage them, then it might be time to seek professional help.

℮ understand your stress

Stress, in association with mental ill-health, is the most common cause of long-term work absence in the UK and many other countries. It's a big deal and is causing huge problems for individuals, the people around them, and society as a whole. We've all experienced it at one time or another, to varying degrees. There are several different types of stress, including time stress (where we don't have enough time to do what is needed), anticipatory stress (worrying about something coming up in the future), encounter stress (worrying about interacting with a particular person or people), and situational stress (experiencing a difficult situation that we may have no control over). Physical stress, from illness or trauma, or stress over past events, are also issues. We may be acutely stressed as we face an intense period of difficulties, or chronically stressed over a long time.

We all know that feeling when our heart beats faster, our breaths become shorter and quicker, there's a knot in the stomach, our muscles tense, palms become clammy and our mouth dries up, and we feel on edge. We might even experience chest pain, dizziness, diarrhoea, or nausea. Internally, a cascade of physiological events is taking place, triggered by how our brain is reacting to an external stressor. Once our brain registers a stressful situation, this fires off an alarm signal to the body, prompting a surge of activity via the sympathetic nervous system.

This system is a network of nerves across the body that responds to stress, putting us on alert. This gives us the ability to respond to perceived danger. Hormones, including adrenaline and cortisol, are released, resulting in physical effects like the heart beating faster and breathing more quickly, to pump more blood and oxygen rapidly to where our body needs it, and also sharpening our senses (such as keener eyesight) to improve our reaction times. Cortisol raises blood sugar levels, boosting the quick availability of energy. Our digestion also slows down to divert energy elsewhere. This system doesn't just kick in during times of mental stress, it also activates when the body is under additional strain, such as when we're ill or doing exercise. Once the perceived threat to the body has passed, things return back to normal. The parasympathetic nervous system sends signals to relax those systems that have been put on alert and brings the body back to its normal functioning after being under stress conditions. Cortisol levels fall, energy is conserved for later use, and functions like digestion pick up again. The parasympathetic nervous system also controls mood and immune responses. So, while the sympathetic nervous system triggers the 'fight or flight' response, the parasympathetic nervous system helps to regulate 'rest and digest' functions.

This dual arrangement is a careful balance of reactions as we require them, and sometimes a certain amount of stress can be helpful. It can sharpen our attention and keep us motivated and focused on a task, and research at University of California Berkeley has found that moderate, short-lived stress (e.g. taking an exam or preparing to give a presentation) can improve cognitive performance and boost memory. However, our fight or flight response evolved when we needed to avoid being eaten by a deadly predator or escape attack from a neighbouring tribe – short-term stressful situations that required a sudden, acute response. Yet, these days, we're more likely to experience stress from multiple non-life-threatening triggers, from major traumatic experiences to minor anxiety-provoking events like running late for an important appointment, the heavy mental burden from financial problems or a high workload, or longer-term life challenges like suffering a toxic work environment, a negative relationship, or relentless caring responsibilities. As a consequence of these manifold encounters, stress and its negative effects build up and this accumulation can make it harder to rest and recover, leading to problems. Stress can make us agitated, overwhelmed, and emotional, to say the least, and can have significant physical costs too.

Cortisol, known as the 'stress hormone', regulates our stress response and has other functions, including regulating blood pressure, metabolism, and blood sugar, suppressing inflammation, and aiding the control of our sleep-wake cycle. Because of the way that our body recognises and responds to cortisol, this hormone can affect almost all organ systems. It's an essential part of our survival but too much of it can be a bad thing. Long-term stress can lead to our bodies being over-exposed to cortisol, which may then disrupt most of our bodily processes. It's no wonder that when we're stressed, we don't feel like ourselves. This disruption leaves us vulnerable to many health problems, like high blood pressure, heart disease, stroke, weight gain, anxiety, depression, memory and concentration issues, sleeping difficulties, and headaches. As you can see, stress significantly impacts the whole body, and being able to cope with life's stressors is not just desirable but crucial.

When we're chronically stressed, it can be increasingly difficult to find the mental and physical reserves we need to cope with the situation. Building our resilience and ability to cope with life's challenges can help us better handle stress, but may be easier said than done, as variations in our genes influence our physical stress response, and our life experiences also play a big part in shaping how we respond. If someone had a significant traumatic experience in childhood, then, as an adult, they may find certain situations trigger an extreme stress reaction. For those who may have been lucky enough to have got through life so far without major traumatic incidents, there will still have been many smaller experiences that influence how they respond to stressors. It could be that interacting with people with certain types of personalities makes us feel tense or uncomfortable (possibly due to previous negative interactions with similar people), or that particular places or situations raise our stress levels (again, due to negative memories from the past). Often, we can't explain why we feel stressed by certain things, but it might be that seemingly insignificant past experiences have laid down memories linked to stress.

It is impossible to live in a world without stressors, so rather than expect to avoid stress we need to find a way to limit the impact it has on us by removing potential triggers, where possible, then working on how we accept and process stress. In other words, it's important to find a way to limit the impact of stress in the first place so we don't become overwhelmed, then deal with it in such a way that we can recover, learn, and move on. It's about being able to recognise what triggers stress in each of us and take control of our response to it, rather than let it control us. It's not straightforward to achieve, of course – nothing is ever easy! We're all so different and what triggers stress in one person might be a source of joy for another. Understanding our own stressors can be an important first step. By knowing what is likely to cause you to feel stressed, you can try to mitigate against these stressors. For instance, if a certain person makes you tense or anxious, you might choose to limit your interactions with them, or the length of interactions, or control what you talk about, try to address head-on what the issue is, or avoid them altogether. The same goes for situations we know are coming up. However, we may not always be able to identify, or predict, our individual stressors. We don't always know what will bother us, why it is causing us to feel anxious, or when something stressful might happen. To deal with the unknown, or those situations we can't control or plan for, we need to find ways of effectively processing our feelings and managing the stress we experience.

Methods to dial down generalised stress involve many of the activities covered across this book – it's essentially the *calmist* approach. From meditation and controlled breathing, hobbies and healthy eating, to avoiding stimulants, practising relaxation techniques, or volunteering, there are many possibilities to try. Much of this is about building resilience by taking care of your rest levels, leaving you with the energy and strategies to cope when life throws trash at you.

stimulate the vagus nerve (or stick to meditation, yoga, or controlled breathing)

At the time of writing, there's an online trend for self-stimulating the vagus nerve to reduce stress and improve mental wellbeing. 'What's that all about?', you might well ask! Let me explain. The vagus nerve, part of the parasympathetic nervous system, is the longest cranial nerve in our body, running from the brain to the intestines. It is crucial in a range of roles, including our immune response, heart rate, mood control, and digestion, and is part of the 'gut-brain axis', which links two-way communication between the brain and the gastrointestinal region. Due to its importance in sending signals up and down the body, and its critical role in the parasympathetic nervous system, being able to stimulate the activity of this nerve may help to boost positive effects, including those involved in our mental state.

Vagus nerve stimulation in a clinical setting is typically carried out by implanting a device under the skin on the chest with wires connecting it to the vagus nerve. When activated, electrical impulses stimulate the nerve to send signals to the brain. This medical intervention is currently used to treat conditions like epilepsy, stroke, and depression that's hard to treat. Regarding depression, while the precise mechanism is unknown, the stimulation is thought to help by influencing specific mood-regulating neurotransmitters (chemical messages) in the brain. There is also a less invasive form of vagus nerve stimulation that involves stimulating the vagus nerve through electrodes clipped onto the ear. Non-invasive vagus nerve stimulation has been a target of research interest in mental health for some time and small studies indicate that it may improve emotion recognition, boost mood, and reduce symptoms of depression. However, the evidence is mixed and the mechanism behind this intervention remains unclear. Deep breathing, yoga, and meditation are also thought to stimulate the vagus nerve but there is little scientific data to confirm this or reveal how this works.

The social media trend hyping self-manipulation of the vagus nerve as a way to induce relaxation and reduce stress has very little to go on in the way of science. Having said that, wellbeing benefits of meditation, controlled breathing and yoga have been identified and it's possible that the vagus nerve is playing a part in this. We just don't know enough about what exactly is happening. Sticking to practices that have been shown to be beneficial, rather than attempting weird and wacky techniques invented by influencers with no good evidence behind them, seems like the sensible option.

☉ cultivate resilience

People who feel more resilient are less likely to be overwhelmed by stress. We can't always control the world around us, so it is important to be able to fall back on our mental reserves to steady us when needed. Some people manage to preserve their wellbeing despite experiencing stress and adversity – they're said to have good emotional resilience. For others, life can get on top of them, and they may buckle when things hit a downturn.

Emotional resilience refers to our ability to respond to stressful situations – how we manage, recover, and learn from them – and how much resilience we have can be due to a host of reasons. Our genes, age, health, and life experiences all shape the extent to which we are resilient, and our resilience levels vary widely. It's not always possible to regulate our response to stress and build resilience. If we're already in difficult circumstances before another stressful event comes along, it can be very hard to manage. For instance, living in poor physical or mental health, being isolated or lacking healthy support networks, feeling unsafe or facing prejudice, living in poverty, or being unable to access appropriate support can all make life especially tough day-to-day. For some people, just getting through one day to the next is challenging enough. While they might have their own coping mechanisms, they may lack the capacity to actively grow resilience. But for many of us, it's possible to find opportunities to cultivate resilience in preparation for the bad days.

Increasing our resilience can be a vital defence against ill-health and indispensable for maintaining our wellbeing, while poor resilience can leave us vulnerable to mental distress and illness. Growing resilience comes from learning how to adapt to your stress and feel more in control when things are tough. Research has found that people who are able to experience some positivity during times of stress, or use positivity to rebound from, or find meaning in, the stressful situation, have greater resilience and are able to more efficiently regulate their emotions.

Building resilience takes ongoing effort. We can try shifting our mindsets to accept what we can and can't control in life, and act accordingly. Learning to better accept difficult situations and people and finding a way to deal with them without impacting your own wellbeing is an important element in building resilience. Unfortunately, this world is full of circumstances that we don't like or aren't fair but we may be able to take control of how we feel about them. We can choose to be angry and fight; we can choose to accept the situation but work calmly to address it; we can choose to drop it. It's about acting consciously rather than allowing ourselves to be consumed by unhelpful thoughts. It may sometimes feel like you're at war with the world, but the best way to avoid this causing complete emotional exhaustion is to pick your battles – try not to fight everything at once; prioritise those that really matter and let other things go.[7] We can also take more time to reflect in the moment and respond to our emotions and needs appropriately. We can increase our energy reserves and practise calming techniques, so we're better prepared for future dips. *calmism* is very much about cultivating resilience – achieving complete rest by adopting regular rest habits to give us a buffer and equip us with the skills we need when times are difficult.

One aspect of resilience is managing our expectations. If we tend to expect the world to be all rainbows and wonderful opportunities, we will regularly be let down. A good example of this is when children are too shielded from disappointment or upset. In this case, they may find it more difficult to cope when something negative happens. After all, not everyone will want to be their friend. They won't be the best at everything. They can't have every toy. They won't win every game. Some honest truths (within reason and stemming from good intentions) can help children to understand that things won't always go their way in life as well as learn how to manage their feelings and responses as a result. In other words, it's helping them to grow their resilience. This is increasingly essential as they grow up. As adults, we're not that much different. We know that life isn't always rosy so we should have realistic expectations of the world, ourselves, and others to avoid constant disappointment and yearning. If we can accept that things won't always be to our liking and that sometimes we, and others, make mistakes – people are not, and are never going to be, perfect – then we may find we're less unsettled by the ebbs and flows of life. This isn't about taking a pessimistic view of the world, or ignoring dreams and ambitions, but recognising that life is complex and ever-changing and sometimes the best thing to do is to roll along with it.

Acceptance of how things are is another feature of mindfulness that aids our inner calm, and the Japanese concept of *wabi-sabi* also inspires us to embrace acceptance. In this philosophy, we are encouraged to accept

---

7   For instance, is it really worth ruining your day being enraged by social media comments from people you don't know and who have no bearing on your life in the real world?!

change and prepare for it as everything is transient and we should appreciate what's here now as it will all be gone eventually. We should accept that while nothing is perfect, we can appreciate it anyway – it is what it is. (I explain more in habit 5.)

⓮ accept who you are

We waste too much time judging and comparing ourselves with others or what society expects of us. We have a tendency to self-criticise, and the external message that we should always be striving to better ourselves implies that we're not good enough now, which doesn't help. We're trying to please others and keep up with expectations and are encouraged to constantly evaluate ourselves and improve. If we're to achieve emotional rest, this way of being has to change. We need to be able to accept, and be comfortable with, who we are and go from there. Self-acceptance is understanding and embracing the whole of ourselves. This doesn't mean we have to be arrogant or naively ignore our less than amazing attributes, and we can still learn and grow, but we should recognise and accept all the facets that make up our complete selves and be comfortable with that.

Studies of university students have found that unconditional self-acceptance reduces emotional problems that hinder educational and personal growth. This is a formative time of life in which they're often still trying to understand who they are, a time when they may doubt themselves or feel self-conscious, and when self-esteem may be easily bruised. It makes sense that those who are more sure of, and more comfortable with, themselves will probably have a less turbulent ride. And this kind of result isn't just found in the young. Self-acceptance has been shown to be a good predictor of wellbeing in older adults too. Self-acceptance is closely linked with general psychological wellbeing, which seems logical – if you're at ease with who you are, you're less likely to be brought down by people around you and less likely to waste time comparing yourself to others.

Instead of shrinking our real selves to somehow fit in, shedding the obligation to meet the expectations of others can be very liberating. It is impossible to please everyone all the time and if you try to do so, you may forget that you have wants and needs too. Although it's important to do things for others some of the time, for true emotional rest it's essential that we each try to live life in our own way to avoid internal struggles over priorities. If we're people-pleasers or have a huge list of responsibilities, it can be hard to move away from a pattern of behaviour that suits other people, but making even small changes can set us on a more positive path. Think about aspects of your life that are either for the benefit of other people or are due to concerns about what people might think of you. Where could you start to break out from your current behaviours? For example, if you always respond to messages quickly but this adds pressure to your day, try letting non-urgent messages wait a little longer and notice whether it really mattered. Or, if you've got stuck in a rut with the way you dress but are worried what people would say if you refreshed your look, try introducing a few small changes (or, if you're feeling brave, go for a full revamp) and see what happens. Chances are that the world won't stop, no one will freak out, and it will all be fine. The main impact will likely be on you and how you feel. A little less frustrated, hopefully.

Clearly, those examples are very small, but they represent a starting point. For some people, a radical transformation is required to achieve how they really want to live, which can be a hugely daunting process but one that is life changing. For instance, individuals who've rejected the life that they felt was expected of them (e.g. choosing a completely different career than expected, leaving a long-term partner, moving to the other side of the world, changing gender) often express this experience as freeing, that they can finally breathe and be themselves, and how it brought huge relief to now live as they please. It's not that life is necessarily going to be any easier, but they feel better now they can live it according to their own rules. Any change to your life that helps you break out of old patterns that you're not happy with can be empowering as you take back some control.

Even if you feel okay about yourself much of the time, something that a lot of us experience now and then is that feeling of imposter syndrome. This is where you doubt your abilities and feel like a fraud, despite being successful and qualified, and you believe that everyone else knows exactly what they're doing and is better than you. This feeling is particularly prevalent amongst high achievers, perfectionists, and those with very high standards and goals, and, in part, reflects a combination of the pressure that we put on ourselves to excel as well as our perceptions of the abilities of others. In other words, we compare our abilities to other people and have false beliefs about what's expected of us, and these imagined perceptions can lead to real consequences. Imposter syndrome is often found in people with depression and anxiety and is also related to diminished job performance and satisfaction and to burnout.

When we experience imposter syndrome, we tend to believe that we're not as intelligent or talented as people may think and that any success we've had was down to luck or the right timing. We feel we don't deserve the opportunities offered to us. This can hold us back. If we think we're going to be exposed as a fraud, we're less likely to take risks, such as going for a job promotion or taking on extra responsibility. We may avoid speaking up in meetings or check our ideas more often with someone else before putting them forward. We increasingly lack trust in ourselves and our judgements. We start to limit ourselves, may live in fear of failure, and self-doubt increases. It's not a good way to live, especially when you only live once. We may know deep down that we are enough, but how can we tell ourselves that?

The first step is to acknowledge that you feel under pressure to perform or behave in a certain way and take time to reflect on why this might be. Looking at everything through a rational lens may help you to gain greater perspective. Question why you doubt yourself – have you any good reason to? Where's the evidence?! It may help to note down your qualifications, skills, awards, and successes etc, so you can see on paper (or hear on a voice note) that you are good enough and have the required attributes to deliver. Talk to people you trust and ask for candid feedback and no doubt they'll tell you that you're amazing or can give examples of how brilliant you are. But you also have to allow yourself to accept commendation and appreciation – it might feel uncomfortable or embarrassing to listen to, but everyone needs to hear it from time to time. Accept that you are deserving of praise and own your successes. It's not always in our blood to boast about our achievements, and we can be much more comfortable talking ourselves down and being critical. We must move beyond this and mute that inner critic. While we don't need to go shouting it from the highest mountain, we can find ways to be proud of ourselves and communicate that when appropriate, to support our self-acceptance.

Of course, we may not always receive praise, which can lead to self-doubt, and this is where we can practise self-praise. Research has found that something called 'reflective recognition' can be a useful approach to improving feelings of pride and empowerment. This technique involves reflecting on and, typically, sharing what you're proud of and why. It is often used in a business context, as part of performance appraisals, but given the beneficial effects, there's no reason why it can't also be employed in a personal context too. Often, we only receive praise on what others have seen or heard about us, but it may be that we want to be appreciated for other things. For example, perhaps you volunteered to help at a community event and people thanked you profusely for giving up your time on the day, but you may have wanted to be appreciated for the creative input you contributed that helped bring the event to life. You know that it was the fact that the event was novel and exciting that made it a success, and that success was largely down to your ideas. When we're not recognised for our whole contribution, it can make us feel under-valued and less likely to make an effort in future – maybe for the next community event you think you might lend a hand for an hour but not bother contributing in any other way because it's not appreciated. Receiving appropriate recognition for our contribution bolsters our pride and helps us to believe in ourselves, but it's not realistic to expect others to offer this all the time. (After all, do you always do this for other people? I doubt it!) So, self-reflection on what we perceive to be our own accomplishments is important and can reassure us of our worth. This isn't about bragging of our brilliance but acknowledging and accepting for ourselves what we've achieved and what led to that, reinforcing our self-worth.

Next, let's quit the perfectionism. As very little in this world is perfect, it means you're constantly striving for something that you're unlikely to reach, so you'll never be satisfied and tend to dwell on mistakes and all the little things that are wrong. For your emotional rest, you need to choose to let some of it go. Stepping back to see the bigger picture can help. Reflect on which aspects of what you're trying to achieve are important and realistic, and what you could learn to be comfortable with if it wasn't perfect. Then prioritise the things that matter. There's no need to argue every point or worry about every detail. Making an allowance for a few things to be just as you desire but finding a way to be more at ease with everything else being good enough will start to relieve the emotional burden.

We can feel better about ourselves and learn to accept who we are with some practice. Try not to be too hard on yourself, forgive yourself, ignore your inner critic, accept you're not perfect but neither is anyone else, get perspective on your strengths, and recognise, but try not to magnify, your limitations. Don't compare yourself with others – no one else is you. Reflect with objectivity and let go of guilt; if you've done something wrong, try to rectify it, apologise, learn from your mistake, and move on. However, often, the guilt isn't well-placed in the first place as it can be over things we weren't responsible for or had no control over. Worry less about other people's opinions of you – accept that they will sometimes have opinions (that's just life), but they're not actually spending as much time thinking about you as you believe they are! Their brains are probably too busy racing with their own thoughts and concerns. As Coco Chanel once said: "I don't care what you think about me. I don't think about you at all."

ⓔ focus on your breathing

One of the quickest ways to bring down the physiological stress response, to gain personal control over a situation, is to calm our breathing. When we feel anxious or scared, this can cause us to feel short of breath and get our hearts racing, but when we're composed we feel the opposite. Although breathing is a necessarily automatic function, we're able to control it to some extent if we wish. But, to understand why and how we control it, let's first take a quick look at how breathing works. Obviously, we know we breathe in air to get oxygen into the body and then breathe out again to remove carbon dioxide, but what else happens?

The autonomic nervous system manages our breathing so we don't have to consciously think about it. Our brain controls how fast or slow we breathe by sensing our body's need for oxygen at any given time and the requirement to expel carbon dioxide. When we're relaxed, we typically breath slowly and evenly, and our oxygen and carbon dioxide levels are appropriately balanced. However, you may have noticed that when someone is particularly stressed or having a panic attack, they take in short, shallow breaths and can go on to hyperventilate (rapid breaths that can leave a person feeling breathless and dizzy). This over-breathing, as it's also known, is when we take in too much oxygen, causing carbon dioxide levels to drop too low. We need a certain amount of carbon dioxide in the body as it plays several important roles, including regulating the pH[8] of the blood and dilating vessels to enable greater blood transportation. If carbon dioxide levels fall, we can experience ill effects, such as difficulty breathing, dizziness, confusion, chest tightness, rapid heartbeat, sweating, visual problems, and muscle cramps. The result can be a vicious circle as experiencing these problems can lead us to over-breathe further and exacerbate the situation. To break the cycle, we need to ensure there is enough carbon dioxide in our system. Sometimes we see people using a paper bag to inhale from and exhale into. This is because the air we breathe out contains more carbon dioxide than the air around us, so re-breathing it in will get carbon dioxide into our lungs more quickly. But learning to slow our breathing and control how we inhale and exhale is also effective and something we should all learn to do even when we're not hyperventilating.

How we breathe has a big impact on how we feel, both physically and mentally. The nature of our breathing has physiological effects on our heart rate, blood pressure, and muscle fitness, for example, but here we're mainly concerned with modifying it to aid rest. Various scientific studies demonstrate that a few minutes of controlled breathing can reduce anxiety, depression, stress, and emotional exhaustion, as well as promote relaxation and improve attention levels. When you're feeling flustered and someone tells you to take a deep breath, there's a good reason for this – it

8   The balance of acids and bases.

introduces a moment of physiological calm, which subsequently helps to encourage psychological benefits. Breathing exercises are thought to help equalise sympathetic and parasympathetic activity and foster a state of physiological balance. Additional controlled breathing may then further enhance parasympathetic activity to aid relaxation and a calm state of being. I can honestly say that a few moments of slow, focused breathing is now my go-to when I feel a bit tense or rushed and it works wonders for me.

There are copious breath control exercises promoted by health professionals, celebrity influencers, and spiritual guides, all suggesting different methods. There are even whole books devoted to the importance of breath control, such is the interest in this area. A well-known name who promotes the critical role of breathing for wellbeing is Wim Hof. He is a Dutch extreme athlete and motivational speaker, known for his records related to cold exposure (think barefoot marathons across ice, swimming in freezing lakes, climbing Mount Kilimanjaro in shorts), who has developed his own particular method to connect us more deeply with ourselves, with others, and with nature. There are three strands to his method, one of which involves highly controlled breathing. He claims that his technique induces a short stress response in the body, which, if undertaken regularly, will result in greater resilience towards everyday stress and feeling more in control. But, be warned – it is pretty extreme! This method isn't for the faint-hearted. In fact, his website warns that the method can affect motor control and even (in rare cases) result in loss of consciousness! He refers to it as controlled hyperventilation as it involves multiple cycles of deep breaths followed by holding the breath for as long as possible. Hof claims this induces a state of calm and bliss, one that is conducive for meditation. He suggests a number of physiological hypotheses for how the technique helps the body and mind but, as yet, there isn't much hard science to back these up. A series of small studies looking into the Wim Hof method have revealed some interesting results, and anecdotal evidence from people who've tried it points towards something happening that may be of benefit, but there's still much we don't know. The method is rather intense for the average person and other, more realistic, techniques shown to have benefits are available.

Some approaches encourage you to alternate your breathing between each nostril, or to hold your breath for several seconds, making certain sounds as you expel air, or holding your lips or another body part in a specific way. However, if you're interested in having a go at controlled breathing (and I recommend trying at least some simple breathing exercises), the following is a typical, and feasible, approach – the aim is slow breathing, not deep breathing:

1. Start by sitting or lying in a comfortable position. (You can try this standing still, if preferred.)
2. Relax your body with your arms away from your side and your legs slightly apart.
3. Focus your attention on your abdomen and inhale slowly, through your nose. Feel your abdomen expand and rise.
4. Exhale slowly through your mouth in a controlled, unforced way. Notice your abdomen lower again. Try to lower your shoulders each time you breathe out.
5. As you breathe in, count slowly from one to five, and do the same (or longer) as you exhale.
6. Breathe regularly and evenly and continue like this for a few minutes.

Even stopping still where you are and breathing in this way for a few moments is beneficial. The aim is to shift your attention from your thoughts and just attend to your breathing and the feel of your body. I often find that I haven't realised how tense my shoulders and body are until I do this and feel muscles start to relax. This technique also ties in with mindfulness practices.

Concentrating on your breathing, particularly when trying to meditate or sustain periods of reflection, can feel stressful for some people at first. You might worry that you're not doing it correctly or try so hard to concentrate that you find you can't get enough breath in. So, if you find it stressful, take it slowly and just try a few breaths at a time. Once you master that, and notice that you feel a little calmer afterwards, try extending the duration of the controlled breathing at your own pace. Research has shown that as little as one minute of focused breathing can help and my own experience shows that controlling just a few breaths can make a positive impact on wellbeing.

One thing the various techniques have in common is that it is recommended to practise them daily to get the most benefit. If we make controlled breathing a habit during regular life, not only will it come to us more easily when we experience times of stress, but it will help us to be in better control of how we feel day-to-day. It is a key tool for building our resilience and managing our emotional rest and is something we can easily do for ourselves, wherever we are.

ⓒ think positively

Are you a glass half-full or glass half-empty person? I must admit that I tend to veer towards the latter of these and have to make an effort to be more cheerful. But that effort could well be worth it as being positive does us good. As well as having better psychological wellbeing and being less likely to suffer from mental ill-health, a positive approach to life has been linked with several other health benefits including lower pain levels and greater resistance to illness, along with greater resilience to life's challenges. A large study of over 70,000 US nurses[9] found that the most optimistic people had a significantly reduced risk of death from cancer, heart disease, stroke, respiratory disease, and infection. Several other studies have found similar results.

Making a link between having a positive outlook and physical health seems curious, and in fact it's not yet clear exactly how positivity improves our health, but something is clearly going on. There are indications to suggest optimists have healthier immune systems and lipid profiles (markers of fat in the blood linked to heart disease risk), lower inflam- mation, and higher antioxidant levels. Healthier lifestyle choices have also been found in optimistic people, such as healthier diets, more physical activity, better sleep, and being less likely to smoke or drink to excess. It's possible that positive people simply live healthier lives (or already have better health) and when they're exposed to illness or increased stress are

9   The Nurses' Health Study – one of the world's largest investigations into risk factors for chronic diseases in women.

better protected in some way, both mentally and physically. On the other hand, negative thinking (such as worrying, ruminating, catastrophising, filtering out the positive and magnifying the negative) is associated with poor health consequences. Negative thinking is coupled with increased stress and is thought to induce raised blood pressure and cortisol levels. Research reviews of many studies consistently indicate that repetitive negative thinking is associated with depression, anxiety, and emotional distress, and has even been linked with more rapid cognitive decline in older people and other adverse medical outcomes, such as poor heart health and immune dysfunction.

When you look to the future, how does it make you feel? Do you feel interested and excited about future plans and possibilities, or do you worry about the worst that might happen and catastrophise about what might go wrong? Psychologists believe that positive future thinking is important for our mental health. If you're able to imagine a positive future, it's thought that you are better able to make plans and goals and develop a purpose in life (I'll talk more about purpose in life in habit 8). Future thinking helps motivate us and reach our long-term objectives. We don't have to think that everything is sunny all the time or bury our head in the sand about problems to experience positive future thinking. It's more about being able to see a way forward or imagine or visualise positive or constructive things happening in the future, such as looking forward to an event or being motivated to work towards a goal. When you can't see anything positive in the future, this can be a feature of mental ill-health, such as depression.

Of course, it would be weird and, frankly, unnatural if we were super positive all the time. Life just isn't like that. Being more optimistic is about choosing to look at things from a more positive point of view, in the context of the turbulence of real life. But some people would have us believe that everything is amazing all the time (I'm looking at you, online influencers!) and this has led to 'toxic positivity'. This persistent image of perfection and an excessively optimistic state regardless of the situation has been found to be harmful. While the term 'toxic positivity' is relatively new and the science in this area only emerging, social pressure to not experience or express negative emotions and wanting to feel happy to an extreme degree is harmful to our mental wellbeing and has been linked to depression. The obsessive pursuit of happiness results in the denial, minimisation, and invalidation of negative emotions and authentic human experiences. In other words, people hide how they really feel, dismiss emotions that aren't positive, may feel shame or failure for feeling negatively, or feel that it's their own fault if they don't feel happy. The preoccupation with a positive outlook, at the expense of any other emotion, can make others feel pressure to be the same and dampen their real emotions (hence the toxicity).

Social media is particularly bad for promoting toxic positivity. It's flooded with images of subjectively perfect looks and lifestyles and meaningless positive affirmations. For every bouncy, hyper-enthusiastic, living-life-to-the-max sort of person throwing out life advice, we need to recognise that what they're choosing to present is not the whole picture and there will inevitably be a lot more going on behind the scenes. They will have good days and bad days, and challenges to their mental wellbeing, just like the rest of us. The bottom line is that we need to find a healthy balance between aiming for a more positive outlook on life and recognising the whole range of our emotions and dealing with them appropriately. One thought-provoking study into attitudes towards life found that rather than seeking pleasant emotions, like happiness, and avoiding unpleasant emotions, like anger, all the time, wellbeing was improved more by wanting to feel more anger or more happiness when these emotions were useful, and less of them when they were not useful – i.e. it was better for people to be able to recognise their true emotions and apply them appropriately in response to a situation. This supports the notion that we need to recognise, understand, utilise, and learn from the full range of our emotions to be able to gain emotional balance.

There are ways to feel more positive without it feeling unnatural. Most of us will have a pretty good sense of whether we are a largely positive or negative type of person (we may, of course, fall somewhere in the middle). If you think you are quite negative but would like to gain some of the health benefits that the optimists are enjoying, the good news is that it is possible to change – albeit with practice. It's about creating new, positive thinking patterns and habits.

It's pretty common to find ourselves stuck in a bad mood. It might be that something triggers this mood but then we expand it to cover lots of other things that are niggling us. Negative thoughts can have a habit of getting stuck in a loop, so we should find a path to break out of this cycle of negative thinking. This isn't the same as simply telling yourself to 'be happy' instead. We should accept that negative thoughts (just like any thoughts) can be fleeting and that we can actively stop the cycle. It doesn't mean that something isn't wrong or that you haven't been made to feel bad by something or someone, for example, but it is about accepting this feeling and actively working to move through it. This is where an approach like mindfulness can be really helpful. It helps us to reflect on how we feel right now and what we can do to feel better in the moment, without worrying about what's next. By calming ourselves in the moment, we gain emotional rest and boost our ability to think more clearly about a situation to be able to handle it better. We can also improve our outlook for the future by identifying things to look forward to, no matter how small, and visualising enjoying them. We can set goals to work towards and try to appreciate the journey along the way by recognising that each step we take is making progress to reaching those goals. Finding an appreciation for the people and things in our lives that we're grateful for is also an important way to gain perspective and feel more positive. Feeling, and expressing, gratitude is a critical component of our wellbeing and I'll talk more about this in habit 8. Other suggestions to increase positive feelings include: prioritise a healthy lifestyle, regularly check in with your thoughts every day to assess how you're feeling and whether you could put a positive spin on any negative thoughts, try to find the humour in everyday situations, let the little niggles go, and surround yourself with positive people (they can help you see the positive side of things and offer support, whereas negative people may increase your stress by amplifying issues). Finding a positive angle might be the difference between thinking 'it's too big a change' and 'it's an opportunity to try something new', or between 'I'm too busy to do that' and 'I'll re-evaluate my schedule and see if I can try something else'. For those in particular need, cognitive behavioural therapy can also be effective.

☻ smile and laugh more

Smiling is another way to feel more positive. It doesn't have to be merely a symptom of happiness but can be used as a deliberate tool to induce positive feelings. Smiling is a universal expression that takes many forms but generally serves a social function conveying a message to the observer. The smile might be real and open but can also be a device to hide fear or be one of malevolence. Most of the time, smiling is something we automatically and unconsciously do when we're happy or amused. Our lips curl up as we flex the muscles at the side of our mouth, our front teeth might be exposed, and the skin at the corners of our eyes may crease. This typically forms a pleasing, friendly expression on our face and it makes us feel good. Yet, despite this being such a basic action and a topic of much research, we still don't know that much about the exact connection between smiling and our health.

Smiling appears to be associated with lower cortisol and increased endorphin levels, and a reduction in stress as well as elevated mood. What is happening physiologically to produce these effects is not fully comprehended, yet studies indicate that smiling is good for us. According to US researcher Marie Cross and colleagues, there is consistent evidence that smiling, whether natural or manipulated (fake) smiles, can have health benefits, including helping to reduce our recovery from stress. There is also some evidence that smiling during physically or socially painful situations might also confer some benefit. And while we may not always be in the mood to smile, it's great to know that pretending to smile can also help to trick the brain to feel better. Researchers have found that body language, including smiling, can influence how we feel, including initiating happiness, so putting on a smile can really make you feel more positive as a result. If you're in a bad mood, trying to smile could help to break the negative thinking. Apparently, the best way to do this is to smile until your cheeks lift and laughter lines start to crinkle – make the smile work its way right up through your face and hold it for as long as you can. It may feel ridiculous but why not try it when you're alone and see how it makes you feel? Even if you feel silly afterwards, that in itself might leave you feeling a little amused and distracted from negative thoughts. Smiling helps to relax facial muscles, so if your face was screwed up tight with stress or anger, a fake smile might at least help to alleviate the tension. I would also add a slow breath or two to assist the relaxation and soften the face.

Smiling is also helpful in connecting with other people and building social bonds. It makes you appear more approachable, and when you smile at someone, they are often likely to smile back at you, amplifying positivity. This visual feedback may make us feel better, help us judge a situation, and aid relationship-building, trust, and engagement. So, even if you don't feel like it, putting on a smile[10] might increase positive experiences and emotions that kick-start genuine smiling and a positive thinking loop.

Just as smiling is good for us, laughter can be even better. Having a good laugh is very satisfying, and listening to, or watching, comedy can be a great mood-boosting experience. Some people giggle away quite easily as part of regular conversation but others may only laugh when they find something really funny. Whatever our style, we often feel good after laughing. (They say that laughter is the best medicine, although if you've broken your leg, for instance, professional medical treatment is probably more effective.) While we've long known about the psychological benefits of laughter, newer research is starting to show that it may also have significant positive physiological effects for people who do it regularly. There are indications that laughter may reduce stress and increase pain tolerance, for example, and a 15-year study of over 50,000 Norwegians found that humour may be a coping mechanism that helps to protect our health, reducing the risk of early death from conditions like heart disease (in women) and infections (in men).

---

10   Just to be clear, I'm not talking about 'putting on a brave face'. That is something that masks your true feelings and may not be that helpful. Putting on a smile is a temporary brain trick that you might use to break out of a bad mood or spark some positive feelings, when needed.

Laughter has been found to increase the release of endorphins (hormones that act as natural painkillers and help to reduce stress and improve mood), relieve our stress response (resulting in a relaxed feeling), and soothe tension in the short term. It can alleviate pain, boost the immune system, improve mood, and help us connect with other people in the longer term. Based on these potential gains, therapists try to stimulate health benefits via the natural, free, and equitably accessible source that is laughter. Laughter therapy aims to help people laugh more easily through laughter-inducing exercises and tools and is often used to reduce stress, provide pain relief, and generally improve wellbeing. Being coaxed to say and do absurd things in the name of humour might sound like your idea of torture but, for even the grouchiest of people, it manages to bring on a smirk or two.

Amazingly, our brains can't seem to tell the difference between natural, spontaneous laughter and fake laughter triggered at will, so just pretending to laugh can also spark health benefits. With no significant side effects, it seems a no-brainer to explore ways to exploit laughter for wellbeing gains. Obviously, it won't always be appropriate. Attempting to prescribe laughter therapy to someone who has suffered a bereavement or has just lost their job could be a bitter pill indeed. However, laughter therapy has been used to positive effect in many groups who are facing difficult circumstances, such as cancer patients and elderly people with depression, and is often used in children's hospitals to help kids with their mood and pain. It seems clear that we could all do more to encourage laughter in our lives, and in those around us, and subsequently enrich our emotional rest. Why not try it… ha ha, hee hee, ho ho?!

what did *the panel* try?

1. Take slow, controlled breaths for one minute each day, followed by asking what you need right now and attending to that. Several participants enjoyed this activity and found it a useful quick break in their day to rest their mind. Some also found it helpful to combine slow breathing with other activities found elsewhere in this book.

I had to set a timer to build this into my day. I often found my mind wandering during it and forgetting about the breathing. However, I stuck at it. Over time, what it's done is make me more aware of my mood. It's easy to rush through your day and not pause to think about your mood and therefore how you're interpreting the things that are happening around you. I'm going to stick at it and keep it going, as I'm improving. Where I do the breathing makes a difference, so I'm trying different places. I've not been so good at addressing my needs on the back of the activity! This is definitely a work in progress! – **Garin.**

I found this quite easy and managed to do it daily. At first, I consciously thought about what I might usefully do next, which led me to read through a couple of magazines that I had been meaning to get round to. I could almost claim that it encouraged me to move on to decluttering some of the stuff that has been hanging around the house – **Paul.**

2. Think about situations where you were concerned about what other people thought of you. Was it worth it, or did it really matter? This activity served as a reminder that spending time focused on what others think about us is unhelpful for our mental wellbeing. It appears to induce negative feelings as most of us are naturally self-critical and we're likely to overestimate the extent to which others are thinking about us.

I tried to think of situations where I was concerned what others thought of me but thinking back to one just made me annoyed. The other one was just minor, and I thought it didn't matter what they thought. I didn't enjoy this activity – **Graham.**

There have been a couple of times relatively recently where I have freelanced for places that I haven't worked for before, and I've left with the feeling that it didn't go that well. Does it matter? Yes, because I might not get invited back, so there's a work/financial impact. Do I particularly care? Yes and no – **Stuart.**

3. Deliberately smile during an activity to encourage positive thoughts. Most participants didn't try this but why not give it a go for yourself and see what happens?!

When I was journaling my achievements, I did feel a bit silly actively smiling when reading the points back and I don't feel it's necessary to do that to get the benefit from the exercise – **Meg.**

4. Increase awareness of your feelings. Set an alarm every day and, when it goes off, stop and reflect on your mood at that point. If it's positive, internally express gratitude for whatever or whoever is contributing to that. If it's less than ideal, consider why, and what you can do straight away to help you feel one per cent better. This simple exercise was universally appreciated by participants. The brief interruption broke individuals out of autopilot mode and gave them greater awareness of, and more control over, their feelings by acting straight away.

Having someone tell me to do this as part of a study helped me do it rather than deciding for myself. It was a great reset and a reminder; an interruption to the day that allowed me to change anything that wasn't working. It allowed me to realise I could change something if I wanted to. If I didn't have an alarm going off, I doubt I would have stopped and evaluated. I have continued to have two reminders on my phone to take a minute each day. I find having the prompt a useful reminder to check in on how I'm feeling – **Katy.**

I found the process easy but I think the trick is putting the time in your phone, so it happens automatically. I like how it reminded me that I was in control. For example, if I was feeling a bit off, what could I do to improve that experience... could it be as simple as getting a glass of water or some mindful breathing? I will continue as I quite like it – **Heather.**

It got me to reflect on my mood for a few minutes, but rather than pause to be in the moment, I took a step to either be thankful (which I found actually lifted an already positive mood) or to do something to lift a more negative mood (such as take a break, have a cup of tea, go outside for some fresh air, change activity). Incidentally, I found that analysing why I felt a particular way did not help. In fact, it was more likely to reinforce those feelings and somehow justify them, like an echo chamber, but doing the positive action was what improved my mood.

And, of course, just the non-judgemental noticing of the mood was actually the first positive step. This activity was also super easy to fit into my day – **Maria.**

I found it really helpful to pause, reflect, and acknowledge the important feedback my mind was communicating, particularly if I was busy in a task or having big feelings, to have the space and permission to communicate with my whole self and to actively listen to what was there. Having the opportunity to celebrate or adjust with a small action, especially if I was feeling a little stuck in my thinking, was healing and a healthy strategy for making a shift in perspective and sometimes my state – **Janine.**

habit 3 :

give your body a break
for *physical* rest

"There is a time for many words, and
there is also a time for sleep"
Homer (*The Odyssey*)

How are you physically feeling right now? Just for a moment, focus on your body. Are you feeling bouncy, raring to go, in tip-top condition, with shiny locks and glowing skin, fit and no health niggles? No? Me neither. In fact, I'm willing to bet that only a tiny fraction of you feel like that, and even then only for a small proportion of the time. The truth is that most of us have times when we just want to stop and lie down. We're worn out. We lack energy and feel sluggish and fatigued, and the signs are starting to show in our bodies as we become increasingly run-down. In a downward spiral, this can affect our mental energy and wellbeing, leading to problems with motivation, emotional stability, and self-esteem, for instance. But the bigger issue is that we also need to keep going – we're busy people with responsibilities and the world isn't going to slow down just for us. So, to avoid burnout and gain physical rest, we need to reclaim control of our energy levels and better manage them day-to-day.

what might help?
ⓔ understand your sleep

No book about rest would be complete without discussion of sleep because when we think about feeling rested, it immediately springs to mind. Most of us would agree that we need more, or, at least, better quality, sleep. We're only too aware of that heavy, foggy and scatter-brained, slow, fatigued feeling hanging over us all day when we've had a bad night's (or many bad nights') sleep. When we sleep well, we feel alert, energised, refreshed, more positive, and better able to function. Sleep is the ultimate tool in supporting our mental wellbeing, but lack of it has negative effects very quickly. After just one bad night's sleep it's more difficult to concentrate, we tend to eat more high-sugar, high-fat foods to give us energy through the day, we may lack motivation, and feel emotionally less stable or simply grumpy. We all know that sleep is vital for our health but, to improve it, it helps to understand what happens when we sleep and why it's so essential for our very survival.

Sleep is a resting state in which your body and mind slow down, unconscious to the world outside. It's a complex process where many bodily functions are suspended until waking once more, while others continue in order to keep us alive, enable repair, replenish energy, consolidate memories, and more. It's a mysterious phenomenon, accounting for a big chunk of our lives (around one third), that we're unaware of during its occurrence and so remember little of. Let's say the average life expectancy is broadly 81 years (yes, I know it's different for men and women and is dependent on where you live etc, but just go with me here); that means we have practically no knowledge or awareness of about 27 years of our life!

There are four main stages of sleep: stage one (shift from awake to asleep) – you doze off and fall asleep; stage two (light sleep) – your body and mind slow down but sleep is light and you can easily be awoken; stage three (deep sleep) – the brain slows further to prevent unwanted awakenings; stage four (REM sleep – rapid eye movement) – the brain's activity increases again to near that of when you're awake and this is when you have the most dreams. During these sleep stages the body also experiences physiological changes, such as in breathing and heart rate, blood pressure, temperature, blood flow to particular tissues, hormones, and digestion. We pass through a cycle of the four sleep stages about every 70 to 120 minutes and, depending on how long we sleep, we complete around four to six cycles each night. We sleep more deeply during the first half of the night but spend more time in the brain-active REM sleep during the latter half, the time when we dream. Of course, many of us don't always complete the four to six sleep cycles as we may have disrupted sleep in the night, or be woken by an alarm or a child, for instance, early in the morning rather than waking naturally. This is one of the reasons why we might still feel fatigued when we rise in the morning, despite seemingly getting enough hours in bed.

It's still not entirely known why we sleep but it appears there may be multiple reasons. At the most basic level, we need sleep to rest our body and mind and conserve and replenish our energy for another day, but there's more going on too. During sleep, the body has the opportunity to carry out restorative activities, such as repairs to tissues, protein synthesis, and immune function. The brain also clears itself of waste products that have built up through the day and there's some reorganisation in the neurons taking place, possibly in response to the brain's daytime experiences. We embed memories during sleep, and sleep relieves stress and depression, as well as sharpens our cognitive abilities. As sleep affects pretty much every part of our body, it also has wide physical health benefits.

While there's an emphasis on sleep duration for good health, it is not just how long we sleep but the quality of it that is important. Repeated and regular disruption to our sleep has consequences. As well as feeling tired, lack of sleep can trigger headaches and migraines. Sleep deficits can accumulate over time and impact our cognitive performance, speed, and accuracy, as well as memory. Fundamental biological processes are interrupted that affect our energy levels, thinking, learning ability, and concentration in the short term, and contribute to other health harms in the longer term. There is a wide range of known and suspected links between poor sleep and health conditions, including hypertension, heart problems, obesity, type-2 diabetes, mood disorders, and dementia. Overeating is also common when we haven't had enough sleep as our body feels the need for more energy. An increased hunger signal from raised ghrelin (hunger hormone) levels leads us to reach for higher calorie foods. One research review analysing multiple studies showed that people who are sleep deprived typically eat an average 385 extra calories per day – that's almost 20% of an average woman's daily dietary intake. Lack of sleep in general affects our psychological wellbeing, disturbing our emotional state, mood, and stress levels. We can become sensitive to very small stressors in a way that we wouldn't be otherwise when we've had plenty of good sleep, and sleep deprivation makes us more susceptible to getting ill, worse at making decisions, and less likely to judge risks appropriately. Research by the University of Surrey found that sleep loss can affect over 700 of our genes, including those that regulate our immune system, our natural body clock, and how we respond to stress.

Sleep deprivation even has some similar effects on cognitive, physiological, and physical functioning as alcohol. Studies indicate that being awake for 17 hours induces the same impairments to function as having a blood alcohol concentration (BAC) of 0.05% (50 milligrams of alcohol per 100 millilitres of blood) and being awake for 24 hours is like having a BAC twice that amount (0.10%). While the legal BAC limit for driving is 0.08% in the US and England, for example, some countries, like Scotland, have a lower limit as driving impairments start to arise at 0.05%. This can inevitably have substantial implications for taking on tasks of responsibility (e.g. medical care, childcare, operating a vehicle or machinery) while lacking sleep. When sleep deprived, we have a higher risk of having an accident, whether on the road, at work or during leisure, or making mistakes that could have serious consequences for other people. Several major disasters have been associated with sleep deprivation. For example, the Chernobyl nuclear disaster was linked to workers having too little sleep, as was the explosion of the space shuttle *Challenger.*

As severe sleep deprivation can lead to such detrimental effects, it is no longer considered ethical to deprive research participants of sleep to any great extent. So, to observe the full consequences we must turn to historical studies (when researchers were less bothered about the ethical implications of their work!). Studies from the past reveal how effects rapidly progress from irritability and anxiety to disorientation, hallucinations, and psychotic symptoms with increasing duration of sleep loss. Some of these symptoms occurred in the studies after just 24 hours of being awake (i.e. one night's loss of sleep), and many were resolved after a period of sleep.

The amount of sleep needed to feel rested and function well varies widely between people. Some need lots of sleep to feel right but others function fine on little sleep. On average, it seems that most adults need somewhere between seven and nine hours' sleep each night to maintain their wellbeing. The amount of sleep we need decreases as we age, and when we reach older age our sleeping patterns tend to change too. Getting the right amount of sleep, undisturbed – or, more accurately, achieving around four to six complete sleep cycles, depending on our individual need – seems to be the key. We can tell if we have good sleep as we're able to fall asleep easily, don't fully wake up in the night, don't wake too early in the morning, and feel refreshed when we get up. To many of us, this sounds like a dream! Our chatty minds prevent us from drifting off at a reasonable time, internal and external influences disrupt our slumber, and we're unnaturally awoken before we're really ready to get up. It's estimated that around six in every ten people around the world feel they don't get enough sleep.[11]

Sleep researchers aren't just interested in the period when we sleep. This is just one part of the sleep-wake cycle we all move through daily – our circadian rhythm. The circadian rhythm refers to the 24-hour internal body clock that regulates cycles of alertness and sleepiness. It is largely affected by changes in light and dark. This rhythm is

---

11   There is such a thing as getting too much sleep, of course. Hypersomnia is a condition of sleeping too much. People with this are often excessively sleepy during the day and find it hard to completely wake up. It is thought to be associated, in part, with an imbalance in the length of the different sleep stages, in particular too little deep sleep at night and too much light, non-REM sleep. Rebalancing sleep patterns is needed to address this.

controlled by a part of the brain's hypothalamus and influences many physiological processes, including sleep, hormones, body temperature, and appetite. The circadian rhythm seemingly evolved to help us adapt to our environment and anticipate changes, such as in temperature and food availability. It enables humans to optimise both energy expenditure and internal physiology. As the day moves to darkness, our brain's pineal gland produces melatonin (sleep hormone) that helps us to naturally transition from wakefulness to sleepiness. As it gets light again, melatonin production stops. As such, melatonin plays a vital role in helping to regulate the circadian rhythm. This is also why light pollution in our rooms (whether from internal or external sources) is detrimental – light exposure at night can inhibit the production of melatonin, making it more difficult to sleep and disturbing the natural sleep-wake cycle. Disturbances in our circadian rhythm can lead to a range of health problems and essentially stem from a misalignment between a person's sleep timeline and the physical and social 24-hour cycle of their environment. If a person is sleeping or awake at the 'wrong' time of day (i.e. when it is either too light or too dark), such as those who do shift work or who can't respond in the same way to light cues, like people who are blind, they are more prone to circadian rhythm disorders. (I explain more about the effect of light on the circadian rhythm in habit 5.) To aid sleep, we need to help our circadian rhythm to flow in tune with the world around us.

When talking about sleep, you often hear people talk about larks and owls – with the larks being the early risers and early to bed, and the owls being the opposite. In reality, there's more variation as many of us don't strictly fit into either category. Various researchers have tried to define possible sleep pattern types but, on an individual level, it's probably more important to reflect on how you personally operate. Not only are we governed by our circadian rhythm, but during each 24-hour cycle there are also multiple, smaller 90–120-minute cycles, called ultradian rhythms, in which we experience peaks and troughs of energy and focus. If you can get the flow of your day to match your natural rhythms, this would go a long way in managing your mental and physical rest levels.

☺ improve your sleep

There's oodles of information available on improving sleep, and if you think you could benefit then do check it out. There's not enough space here to cover it all but I'll summarise some of the advice and ideas that you might try (many of which may be obvious but are worth reminding ourselves of). Bear in mind, however, that there's no single solution as we're all so different in our sleep needs and personal circumstances. (There's more detail on many of the points below throughout the book.)

• *Establish a healthy sleep routine, if possible* – go to bed at a time that would allow for seven to nine hours of sleep, depending on your need (e.g. if you need eight hours sleep to feel rested and you tend to get woken up around 7 a.m., then aim to be asleep by 11 p.m.); go to bed, and get up, at the same time each day to train your body and mind into a healthy routine; listen to your body – if you're tired, go to bed, don't fight it to finish watching a movie or send a few emails, for instance.

• *Facilitate the right state of mind to ease into sleep* – try mindfulness, controlled breathing, journaling, listening to calming music, reading a book, having a warm bath, a relaxing hobby, some quiet time, or other soothing techniques to still the mind before hitting the hay; avoid television, screens or doing work, which tend to stimulate the brain, for at least one hour before bed.

• *Ready your body for relaxation* – get fresh air and exercise every day; make sure you eat enough so hunger doesn't wake you but avoid foods and drinks that may interrupt sleep (e.g. caffeine, alcohol) or cause digestive discomfort; avoid exercise and eating in the hours before bed, although gentle stretching combined with controlled breathing and meditation may help; dress in comfortable sleep wear (clothing that causes you to be too hot or cold, itchy, or presses against your bladder, for instance, is not going to help).

• *Create a sleep sanctuary* – do what you can to produce a space that is conducive to sleep: dark, cool, quiet, and comfortable; add small comforts to induce a tranquil atmosphere (e.g. fresh, soft bedding, a calming aroma, tranquil pictures, soft paint colours – even your favourite teddy bear, if you so desire); keep the space tidy to avoid the stress and distraction of clutter.

• *Minimise interruptions* – where it's not always possible to achieve the perfect sleep conditions, try to limit issues: wear an eye mask and/or ear plugs; avoid technology alerts sounding in the night (e.g. phone, washing machine, dishwasher) by using 'do not disturb' features; avoid running machines during your sleep hours, or switch them off altogether; avoid light exposure that may disturb the circadian rhythm (e.g. light from screens like a phone, laptop, or the standby light on a TV) by putting items out of sight – there's a good reason why health professionals often suggest you move the TV out of your bedroom; place your mobile phone out, or at least on the other side, of your bedroom so it's not the first thing you reach for when you can't sleep.

Thinking about what your perfect sleep and wake times would be, and when you're most physically active or mentally alert, can help you to better plan your routines to capitalise on your natural energy levels. I'm hopeless very early in the morning, which is frustrating for my family as they are early birds, and taxing for me as they demand my attention or input when my brain isn't fully awake. This situation isn't ideal for my rest levels. While there isn't much I can do to change that, other than for us to try and work around each other's needs, I can better plan the rest of my day to suit my brain activity. For instance, with my work, the best time to do focused research is mid-morning when the analytical part of my brain seems to be most alert, whereas my writing flows quickest late afternoon and early evening (it's very inconvenient to have to stop writing to make dinner!). For an hour or so after lunch, my mind tends to be sluggish so that's when I get up and out for a walk or do something physical that doesn't require too much concentration. (I fully expect that when I'm retired this is the time of day I will naturally drift off into a nap, only to wake around 3.30 p.m. and then not be able to get to sleep at night, thereby perpetuating the issue.) Consider when your brain and body seem to work best throughout different points in the day. You can exploit this knowledge to your advantage, planning your activities to match your energy levels. This is an efficient way to channel your energy and be productive while limiting exhaustion. Undoubtedly, there are many factors that impinge upon when we can do things – like demands from work, family, or other commitments – but there will be some things you can change, and it may be possible to talk to your colleagues, family, or friends, for example, about moving some things around to help you do them more effectively.

Another tool for topping up sleep is to fit in a quick nap; not always a very practical option but many people rely on them. Daytime napping is common, especially in countries where a siesta is a part of daily life. Socially, while some see this extra snooze as necessary for optimal performance, others view it as being lazy. Our individual circumstances (e.g. older people, shift workers, people with disturbed night-time sleep, those with health problems) and the society in which we live (i.e. whether it is viewed as acceptable or not) typically influence whether we need or choose to nap. Some people are very efficient nappers, managing to drop off quickly, snooze, then wake up reinvigorated after 20 minutes or so. It's like they're plugging in to recharge their batteries. For others, a nap can leave the brain sluggish and affect the following night's sleep.

Scientists have looked into the effect of naps and discovered that they can improve our cognitive performance, stimulate creativity, and improve memory. This is great news for those who crave a few zzzs and can sneak them in. The benefits seem to be most apparent between 30 minutes and two hours after the nap, and the nap should generally be taken in the early afternoon. Such is the power of the daytime nap that some large companies around the world provide rest and sleep areas in the workplace for this purpose. The optimal duration for napping is suggested to be around 20 minutes but no more. Firstly, this is to limit any upset to night-time sleeping, but a bigger reason is that after about 30 minutes of napping, the brain falls more deeply into sleep, leading to something called sleep inertia. This is the transition from the sleeping state to the waking state, during which the ability to think and perform is reduced as we feel groggy and unfocused. This is why beneficial effects of napping don't really kick in until after around 30 minutes of waking from the nap. Sleep inertia is greater following longer naps, so the advice is to keep it short. Most of us don't have the time in the day to nap but if we could find the time, we would probably need to get it done swiftly to then carry on with our other tasks and responsibilities.

To most efficiently nap, we need to be able to drop off quickly and this is something that a lot of us find tricky. However, it is possible to train ourselves to do this, as evidenced by elite athletes who have mastered this skill. Power naps are an essential part of elite athletes' routines for optimal sleep benefits, recovery time, and peak performance, especially when many get up very early for training. Naps are prioritised by most elite sportspeople, and huge investment, in both time and money, is poured into understanding and facilitating their individual sleep needs (such as creating the right sleep environments, providing bespoke bedding, and personalising sleep routines). Athletes aren't any more tired in the day than your average person, but they are very successful nappers because they've learnt how to sleep pretty much at will. They have to be able to sleep quickly, in unfamiliar environments when travelling, and switch off despite their busy schedules. Unfortunately, it's tricky to get information on exactly how they do this. It seems that many keep their strategies secret to maintain a competitive advantage. However, we do know that they use techniques like mindfulness to quiet the racing mind before a nap and try to create the right conditions for sleep.

To help us fall asleep quickly, there are a few things we can try. A few minutes of controlled breathing and mindful meditation can help calm the mind in preparation for rest. Trying to nod off on the sofa with the curtains open and TV on may prove more difficult than if we go to bed, with the curtains closed and an eye mask and ear plugs to shut out light and noise. We can also make the most of the body's cues for taking a rest. After lunch, many of us naturally have a post-prandial drop in energy. This is the ideal time to head for bed. Setting an alarm reduces any worry about oversleeping, and trying to mitigate against anything that might interrupt your sleep can also help (e.g. let others know you're going for a nap and not to disturb you).

Such is the pervasiveness of poor sleep, there is no end to the plethora of gadgets and products aimed at improving our slumber, with some more helpful than others. One piece of kit that might prove useful is the wearable tech that many people already have. Watches that count our steps, for instance, often also measure how we sleep. While they might not be highly accurate (they generally guesstimate how long you're sleeping, for instance), they can provide a broad picture of how you're sleeping and may offer clues as to why you're feeling extra tired one day or more refreshed on others. They can provide detail on not just how long you slept for, but how much time you spent in deep sleep or light sleep and how many times you woke (even though you may not remember this). This information is useful for understanding the quality of your sleep as well as the quantity. As the technology advances, sleep labs could become a thing of the past as clinicians will be able to monitor people's sleep in their own natural environments for more accurate analyses of their behaviours and problems.

There are also gadgets that make sleep-inducing sounds or offer guided meditations, ones that provide soft, gradual light to help you sleep and/or wake more naturally, some that regulate air quality, and products that enhance comfort, such as weighted blankets. Weighted blankets have been popular in recent years and aim to provide a form of pressure therapy whereby the weight is supposed to make you feel secure, as though being hugged or swaddled. This is designed to be calming and aid sleep. Just like any sleep product, they're not for everyone, and it is recommended that some people don't use them (e.g. young children, people with breathing conditions or claustrophobia). I once overheard a conversation between two women discussing weighted blankets. One said, "I think you either love them or hate them," to which the other replied, "I tried one and it made me feel trapped and panicky, sort of like I was lying under the weight of a large, dead animal." I'm guessing she didn't like it! Despite their wide availability, there is little evidence to show that weighted blankets improve sleep for most people. And the same goes for a lot of other sleep gadgets and apps on the market. The usual advice is that if you already find them helpful and not harmful, then that's great and there's no need to change. But if you were thinking of buying one, maybe see if you can 'try before you buy' – borrow one from a friend to find out if you like it before investing in something that may be ineffectual, or even cause problems.

℮ physically rest and relax

We all have those times when we find ourselves staring out of the window but not really looking at anything, or trying to read a book but the words aren't going in. Our mind keeps wandering and we're not really focused on anything. We're daydreaming. It's not something we do on purpose; it just happens. It's as though the brain is taking over and doing what it wants and needs – taking some downtime.

Sleep experts are now recognising the vital importance of what they term 'waking rest' – a period of quiet, reflective thought that gives the brain time to process whatever arises spontaneously. This is needed for at least a few minutes, several times each day. It's like providing a comfortable, empty space, where there are no interruptions or distractions, in which the mind can roam without effort. Waking rest occurs when people relax. Thoughts will pop up, come and go, and meander without direction. People who see you might comment that you seem 'lost in thought' or 'miles away'. You don't need to be sitting or lying still for this. You could be doing something repetitive that takes little concentration, such as going for a walk on a familiar route or doing the washing up.

Investigations into waking rest are relatively few to date, but the evidence so far points to there being cognitive benefits of allowing the brain time to process and consolidate information. Similar neurophysiological rhythms that are found in sleep are also seen in waking rest. Amanda Lamp and colleagues from Washington State University believe that creating time for waking rest may be particularly important for our feelings and emotional control. This is something we can benefit from daily and may be especially important for mental rejuvenation when we're lacking sleep. Other research has found that waking rest may help consolidate memories. After a period of learning any new information, we may need wakeful rest for the brain to imprint it. In fact, even a few minutes rest with your eyes closed can improve memory. Perhaps this is one reason why after a particularly busy day, when asked what we've been up to, we often can't remember. What may seem like a waste of time to some, waking rest or daydreaming could be an essential factor in forming long-term memories.

There's a suggestion that if we don't allow ourselves time in the day to seemingly do nothing and zone out, this may make it harder for the brain to switch off when we need it to most – i.e. when we need to sleep. We're simply too 'on'. For example, when we have a spare moment, like standing in a queue or waiting at the school gate, there's a tendency to check our phone, switch our headphones on, or find another activity to fill the time. Instead, it may be better to do nothing, be in the moment, and allow the brain to naturally switch in and out of focus. Giving your brain some space in the day may improve its ability to better clear itself of thoughts when it's time to wind down and sleep and recoup some rest.

It's not just our minds that need to switch off, of course. The body is good at telling us to stop and rest when it needs it (whether we listen to it or not is another matter). We get slower, lack energy, feel heavy, and just need a sit- or lie-down. We only have so much energy to keep going and our body can't continue to function at full speed. It needs time to replenish its energy, grow and repair, and build strength and stamina, for example, and without rest the body is more vulnerable to injury and illness and takes longer to recover. Even the super fit, who train like demons, know the importance of rest days for recovery. So, it's not just okay to take a break, but essential. How much physical rest we need depends on many elements like our health, age, genetics, fitness, and current activity level. The key is to really listen to your body and not ignore the signs that are calling for rest. There are also suggestions that we should slow down to improve our mental wellbeing. While scientific evidence for a specific link between moving slowly and mental health is limited, the general idea taps into the need for us to sometimes take things more gently, tune in to our body and mind, and notice the world around us. Instead of rushing around all the time, a calmer, more mindful approach can help us to feel more rested and less stressed. You need to remind yourself that you're not being lazy or doing nothing, you need to rest (now and then) to keep your body and mind healthy.

Clearly, we can't lie back and rest all the time. That wouldn't be good for us either. Humans have become increasingly sedentary as techno-logical advances have meant we barely have to move to get food, do work, or seek entertainment. Many of our requirements can be satisfied at the mere touch of a button so we've become an idle lot, fuelling a huge public health problem. Prolonged inactivity is a major global cause of ill-health. We have to purposefully move our bodies, not only to maintain our health but also to experience satisfying physical rest.

ⓔ move more

Running, swimming, dancing, football, gardening, walking, trampolining, caber tossing, cheese rolling… whatever your activity preference, the most important thing is to just do it. Even though we don't always feel like doing exercise, it generally makes us feel better. We seem to be more attuned to our physical self as the blood courses through our system after a workout. Along with the well-known physical benefits, exercise is great for the brain too. It has been consistently found to decrease stress levels, enhance mood, and improve cognitive functions such as memory, focus, and self-control. Other benefits include better sleep, increased energy and motivation, and reduced fatigue. Regular exercise helps to keep our tissues, muscles, and circulation healthy over time, as well as reduce symptoms of anxiety and depression. It can even produce new brain cells.

It may sound counter-intuitive, but exercise can be relaxing. We tend to sleep better when we're physically tired rather than mentally tired and, while it might seem like a contradiction, using our energy up physically can help us to renew and replenish our energy, as well as build stamina. Physiological changes take place as you exercise that both stimulate and calm, which help to alleviate stress. While many claim that exercise promotes a release of endorphins that make you feel good, some researchers are sceptical and believe that the relaxing effects are due to other biochemical substances. David Linden, from Johns Hopkins University, suggests that, rather than endorphins, endocannabinoids may be responsible for improving mood and inducing relaxing sensations after exercise. Whatever the cause, it's clear that exercise helps mental wellbeing, and it's no surprise that it's frequently recommended by mental health professionals. On top of the health effects, you're also likely to feel pleased that you've done some exercise, knowing that it is good for you.

We don't always have to get our heart racing to gain benefit from movement. The physical act of stretching also supports our wellbeing. After sitting for hours at a desk hunched over a keyboard, or a night's sleep in an awkward position, a good stretch makes us feel a lot better. Stretching increases blood flow and relaxes tense muscles, helping to keep them flexible and healthy, aiding physical performance, and decreasing the risk of injury. By releasing some of the tension in our body, stretching starts to reduce physical feelings of stress but research shows that it may also have benefits for the brain. There are indications that stretching may increase serotonin levels (a neurotransmitter that helps us to feel calm and stable) and release endorphins. If we feel good after simple stretches, by adding in mindful and meditative elements we may generate further gains. The most common way to do this is through yoga.

Yoga is much more than doing a headstand on a cliff at sunrise, wearing stretch leggings and a sports bra. While such enviable health and fitness can be the result of regular yoga practice, the aim of yoga is more extensive. In the purist forms of yoga, the aim is to achieve inner well-being and harmonise yourself with the universe. Yoga's roots hark back to ancient India, as a blend of practices including mantras and rituals, but it has gone through several transformations in subsequent centuries.

In Sanskrit, yoga means 'to yoke' or 'to unite' and is widely considered to be the union of the mind, body, and breath. Traditional yoga involved eight core elements, which included elements like restraint and moral discipline and positive duties and observances, alongside posture, focus, and meditation. Typically, however, modern yoga classes only use a select blend of the elements to incorporate balance, stretch poses, breathing, and meditation to centre the body and mind.

Contemporary yoga encompasses a range of different schools, each with their own principles and styles of practice. Styles you may have heard of include Hatha (a bit of a catch-all for the physical side of yoga, an entry point with a classic approach to breathing, postures, and meditation); Ashtanga (a physically demanding series of postures, with spiritual elements); Bikram (a form of hot yoga, taking place in a sauna-like room, with strict rules); Vinyasa (poses synchronised with the breath in a continuous flow);

Kundalini (combining spiritual and physical elements, with an emphasis on releasing kundalini energy said to be trapped in the lower spine); power yoga (an active form, with a quick pace, focused on the workout); and restorative yoga (focusing on relaxation of the body and mind, with very slow poses, meditation, and breathing control). There are many more styles, plus blends with other fitness techniques and modern takes. There's also something called yoga nidra. Rather than a type of physical yoga, it is an element that can be included within other styles. Yoga nidra is a form of guided meditation that is said to tap into a state of relaxed consciousness, leaving the mind somewhere between wakefulness and sleep.

Yoga has long been believed to be beneficial for our health, and science now backs this up. One scientific review of eleven studies that looked at the effects of yoga on the brain found a positive effect of yoga practice on several brain regions, including the hippocampus, amygdala, prefrontal cortex, cingulate cortex, and brain networks. Changes seen in these areas are particularly strong in expert yoga practitioners, resulting from regular practice. Evidence of brain changes is always intriguing but the interpretation of what such changes means in effect is still unclear as the results vary widely across studies. One interpretation is that an increase in hippocampal volume following frequent yoga practice may drive improvements in learning and memory. Similar findings have been seen in studies investigating the impacts of mindfulness and aerobic exercise, indicating that similar benefits might be derived from any, or all, of these activities.

To find out more about the restful effects of yoga, I spoke to yoga teacher Heather Pearson, who is also a public health expert. She explained what the essence of yoga is and how it informs her practice.

Many people feel they can only do yoga if they go to a class but actually it's something that can be done in a number of different ways, anywhere, as long as the connection between the mind, body, and breath is happening. When I was young, I was attracted to the power yoga scene, in a fitness way, and wanted to look like Madonna, but my yoga practice since then has changed leaps and bounds. There's so much more to yoga than an intense physical workout. My yoga has become much slower. I've learnt much more about restorative yoga and seen the value of being still. It calms you and allows the body to rest. I strongly believe that we don't have enough of this in the world today and if we all did restorative yoga, we would be in a much better place.

I asked Heather how this very still form of yoga was different to just lying on a bed, having a rest. She likens yoga to a physical mindfulness practice in that the connection between the mind, body, and breath are at its core. It is in this way that it differs from activities like Pilates or simply stretching.

I meet people all the time who say that they just want to do strong yoga practices, to get in shape, but I find that those are the people who really need to calm down and do the chilled practices. I now typically work with older athletes who were sporty their whole lives but now they're noticing their bodies are really tight and they're getting injured more often, which is why they come to me. I have one particular client with a lot of back pain and we've been doing a lot of mindfulness practice and breathing within our yoga sessions. He says that it has transformed his life – the back pain has significantly reduced and when he does the mindfulness and breathing it is often the only time he feels no pain. He said he's never taken time to sit and think about the way his body is and how it feels – he's always just gone through the motions. It's like there's this light bulb that's been turned on.

I think this last point is salient for many of us – how often do we spend time really focusing on our bodies and how each bit is feeling? We probably only notice when we have some sort of pain. Heather has also witnessed profound effects of yoga practice in people with extreme life circumstances.

Years ago, I helped teach yoga to a group of young people in Sierra Leone – many of whom were former child combatants and members of gangs. These were people who had a lot of external pressures and coming to the yoga mat gave them the opportunity to just focus on the present moment. I saw that having this physical practice, moving their bodies in new ways, and breathing through their movements was really beneficial. It was such a positive way to channel their energy. A couple of the students eventually trained to be yoga instructors themselves and then fundraised to establish their own studio. It was incredible.

Wow! That is quite the advert for yoga!

Moving more and in the right way, whether by stretching, doing yoga, or another activity, could help tackle one bad habit that many of us are guilty of. Poor posture is a big contributor to our mental and physical strain. Our typical lifestyles have led us to this undesirable state of affairs. Sitting for hours at a computer, driving long distances in a slouched position, carrying a toddler on the same hip or a heavy bag on one shoulder, slumping on the sofa when we're tired. Whatever the cause, bad posture can result in tension in the neck, back, and shoulders. As well as headaches, this can lead to increased stress, and stress itself can also result in tension leading to a cycle of further stress. Sitting or walking upright, with good posture, can help to stimulate positive mood and reduce fatigue, and may be a proactive tool we can employ to build our resilience to stress. Improving posture is one simple thing we can all try to lessen feelings of stress and fatigue and increase a sense of physical rest.

what did *the panel* try?

1. Set a healthy sleep routine (go to bed before 11 p.m., get up at the same time each day, no screen time in the hour before bed, no food or drink in the two hours before bed). This activity took a lot of effort for some participants – it can be hard to break out of old routines and set new ones – but some impact was felt by those who managed it.

This was a mixed experience, mostly due to having a small child that makes waking up at the same time very difficult! The requirement to not eat before bed was interesting; it made me realise I quite often nibble a lot (sometimes needlessly) during the evening, and I felt much more comfortable at bedtime for not doing that. I LOVED having no screens before bed. I tend to end up watching TV because that's my partner's preference for winding down and then I get drawn into endless life admin. Having a reason to not get pulled into a list of to-dos all night, and an excuse to just have music on while I pottered with hobbies I always think I don't have time for, was really relaxing and much more enjoyable. I can see how over time it would have a positive influence on the quality of my sleep. I've had periods of time in the past where I've spent most weekday evenings doing something like crafts or a jigsaw with no screen time and now recall how content and relaxed I felt for it – **Caroline.**

The earlier my bedtime is, my ability to wake up in the morning is much better (no surprise). I am going to try and continue going to bed earlier, as it is some-thing I'd like to stick to. However, the no screen time before bed has been an utter failure. I've become much more aware of how much I use my phone in the evening – not just idle scrolling, but it seems that I leave a lot of phone activities for the evening! Duolingo, Noom, responding to people's messages... I'm now trying to figure out why I put them off during the day – **Debbie.**

I know that when I prioritise more sleep, I always feel better for it. Just having the discipline to put enough time aside to sleep is tricky. As I had Covid and my main symptom was tiredness, I am not sure on the impact of this activity on this particular week. I probably would have slept more anyway. However, it helped to have a good routine for the week, and one that I could stick to as my husband was away and I need more sleep than him! – **Liz.**

One volunteer didn't need to be assigned this activity as she reported an enviable talent for sleeping.

I always do good sleep hygiene. I'm a non-functioning wreck unless I get to sleep early, wake up early, don't eat late etc. I'd go so far as to say I'm a champion sleeper. It's probably my greatest skill/only hobby! – **Jenny.**

2. Stretch for 10 minutes each day. Panel members were sent a simple set of stretches to do at any time before 6.30 p.m. Finding the time to do the stretches was a hindrance for a few participants but those who managed to fit them in noticed immediate benefits.

When I was getting bored with work (working from home) and not feeling very motivated, that's when I tried the stretches. The few minutes away from my desk to get my body moving, and stretch out from having been sitting down, did help me focus more on work afterwards – **Julia.**

I did two sets of stretching during my early morning walk with my dog. The stretching made me feel relaxed but energised at the same time, ready for my day – **Marta.**

I did the stretches one morning in my bedroom, though my wife was in bed at the time and asked what I was doing and if I was winding myself! I feel unfit, so this was a good start along the path of doing something and following guidance on specific stretches. It felt good. I would like to continue with these stretches and, better yet, to do them in addition to more regular running/walking, as I've done in the past. Many years ago, I used to run marathons and ultra-marathons, but I haven't done so for a long while as it's hard to fit it in – **Duncan.**

3. Try yoga. This wasn't a specific activity that was requested but some of the volunteers have been practising yoga for a few years to feel healthier and more grounded.

When doing yoga – especially outside – I feel very connected to the environment and can take respite from everyday concerns and worries. Having an in-person class does this in a way an online class doesn't. Somehow the combination of smooth-paced and predictable physical exercise and mental simplicity provides calming and solace from life's problems. Getting better at something, from no matter how low a level, is very affirming – **Graham.**

I started off yoga in my usual fashion, which was to try to be really good at it really quickly, do it every day, and concentrate on power yoga. I really hammered my body trying to get strong and do the big moves. All a bit too much for a 48-year-old! Then I realised that this wasn't being very kind to myself and wasn't doing me much good and I discovered some of the slower stretches and more relaxing elements. I try to focus on getting out of my head and into my body while I'm doing the stretches. I think I'm stretching and relaxing but if I really focus on softening everything, I generally realise that I'm still tense and can let go more. So, it's become a way of looking after myself now rather than trying to be like She-Ra! – **Louise.**

4. Perform regular physical activity. Several participants already incorporate exercise routinely into their week and do so because it makes them feel good both mentally and physically.

I run every other day and, on the days I don't run, I do a series of exercise videos. In general, I like to exercise regularly because I enjoy the feeling of having worked hard and having completed a workout. I want to be physically fit and strong, and I also think of it as my 'me time'. It's something that I do entirely by myself, for myself, and I enjoy that (with the benefit that it doesn't make me feel guilty at all because it's healthy and something I should be doing). I've been running at least this much for the past 20 years and I've continued because I love it. It calms me and makes me happier throughout the entire day. I often say that running is my therapy. I find it provides me with peaceful moments where I can zone out or, more often, think through an issue. Whether it's a huge life problem or something small like what to buy someone as a gift, exercising helps to make the answer clearer for me and sort out my thoughts. When at a crossroad in life, most people will say 'I need to sleep on it', but I'll say 'I need to run on it'. People who live with me will readily tell you that on the occasions where I haven't been able to run due to injury, I am a lot moodier and more difficult to be around! – **Meg.**

habit 4:

nourish from within
for *nutritional* rest

"One cannot think well, love well, sleep
well, if one has not dined well"
Virginia Woolf (*A Room of One's Own*)

Let's face it, we're all on an eternal quest to find hacks and shortcuts to make life easier, smoother and less stressful. A typical place we do this is in our diet. I think it's fair to say that not many people prepare wholesome meals and snacks all day, every day, all year round. Even the most motivated and well-intentioned individuals seek the ease of quick, convenient foods and drinks, plus the comfort of treats we struggle to deny ourselves, when we're having an off-day or want to celebrate – the pick-me-ups. I'm looking at you, chocolate, ice cream, sweets, cake, crisps, beer, wine etc. A treat now and then isn't a big deal but many of us get into bad habits. One cookie leads to another, one beer turns into three, and a quiet night in turns into a fast food fest! The problem is that there are often consequences to pay. It's widely known that too much of a good thing can contribute to significant health problems, from weight issues to heart disease to liver problems. But even small deviations from a healthy diet can influence how we feel.

We often behave as if it's a numbers game – we need to consume enough calories to fuel our physical needs to keep us going; too few and we lack the energy we need, too many and we may gain weight. But food and drink are so much more than just fuel. Along with energy, we consume nutrients, additives, and a whole array of other compounds, and each has its own role and effect on the body. There is a strong link between what we eat and drink and how we feel. The physical effects of our diet might be very obvious (e.g. energy levels, weight gain, allergies), but changes are also happening at the cellular level and there are effects on the brain and our emotions. Eating well can help us grow, repair, and stay healthy, and also boost wellbeing. What we put into our bodies has such profound effects that by paying careful attention to what we consume, we can reap significant rest rewards.

It's harder to enjoy life when you don't feel well inside. I'm not talking about being ill (which will clearly have a big impact), I'm referring to simply feeling a bit off, not quite yourself, a bit sluggish or bloated, or lacking your usual energy. Often, these symptoms are down to what we're eating and drinking, and what we need is to give our body a break – in other words, nutritional rest. Making improvements to what, how, and when we consume could help us feel a lot better and calmer inside.

what might help?
℮ eat enough

The human brain uses around 20% of the body's energy and, as it plays a critical role in how we feel, we need to make sure we suitably fuel it. Our energy comes mainly from glucose, a simple sugar. The body turns the foods we've eaten into glucose, which is then absorbed in the intestines and passes into the bloodstream. This is what we're referring to when talking about blood glucose, or blood sugar, levels. Most of the cells in our body need glucose to function properly, including the brain, so maintaining consistent levels is essential. If the brain doesn't get enough glucose, it won't have enough energy to ensure effective communication between its neurons (nerve cells) and the rest of the body. Neurons trans-mit information both within the brain and around the body, so anything that interferes with their performance also interferes with the signals being sent. In the short term, a lack of glucose, or energy, to the brain can leave us tired and irritable and we may not be able to concentrate or remember things. You'll recognise these symptoms from times when you've missed a meal or not eaten enough in a day. Skipping or having irregular meals, doing intensive exercise, or being on a restrictive diet can all lead to an energy deficit. Longer term, the effects are more serious, including cognitive impairments. So, to feel right and avoid energy highs and lows, we need to make sure we eat enough in the first instance.

℮ but don't overdo it

Of course, many of us consume too much sugar and this isn't doing us any favours. While a deficit might impact our cognitive functions and emotional state, consuming extra glucose beyond the normal range doesn't help improve our performance or memory in any way. Any glucose that's not needed for healthy bodily operations is stored for later use. We get much of our glucose from the carbohydrates we eat, such as bread, rice, potatoes, cereals, fruit, vegetables, and sugars. Although glucose is a form of sugar, we don't need to consume lots of sugar to acquire what the body needs. It's tempting to fool ourselves that eating a bar of chocolate is giving our brain energy to concentrate, but we can get plenty of glucose from healthy, non-sugary foods.

When tired, we often turn to sugary foods and drinks for a rapid energy boost. To get a quick fix, the brain urges us to consume something sweet that can release energy fast. Eating sugar fires off a surge of dopamine, a chemical message (or neurotransmitter) that is sent from a neuron to a target cell. Dopamine is involved in various functions in the body but is best known for making us feel good, helping us experience pleasure and satisfaction, and making us feel motivated. It's like a reward and is implicated in why we crave pleasant experiences, including eating tasty food. Dopamine also plays a role in our mood, sleep, ability to concentrate, memory, and learning. The reward of pleasure stemming from the consumption of sugary foods becomes reinforced in the brain. The more we experience the pleasure, the more we want it. This is why dopamine is a major chemical involved in addiction. So, in theory, the more we give in to temptation and have that refreshing, sugary drink or appealing cake, the harder it will be to resist in future. And it often pans out that way.

We don't just turn to sugary foods when we need energy; we often reach for them when our mood is low, and that's because our brain is telling us that they will make us feel better. Comfort eating is like a psychological sticking plaster. The brain thinks it will get a dopamine hit from the sugar that will subsequently increase the sensation of pleasure and lift our spirits. But any effect is short term. It doesn't last. And, of course, it doesn't always work. A review, by scientists at Humboldt University and the Universities of Warwick and Lancaster, of 31 studies found no positive effect of carbohydrate consumption on any aspect of mood. These researchers, and others who have found similar results, are not convinced that sugar-rich foods improve our mood at all. They did reveal another finding, however. While moods were not lifted, carbohydrate consumption was linked with increased fatigue and reduced alertness in the first hour after consumption. That may be because consuming sugar doesn't just make us want it more; it also messes with our blood sugar levels.

As glucose enters the bloodstream after eating, blood sugar levels start to rise. This prompts the pancreas to produce insulin, a hormone that helps transfer the glucose to your cells for energy, either for use straight away or stored for later need. As cells absorb this glucose, the level of sugar in the bloodstream begins to fall back to the level it was before you ate. In a healthy person, blood sugar levels rise and fall several times throughout the day. Foods that take a long time to break down result in a gradual release of glucose into the blood and a measured effect – from the rise in the bloodstream and the uptake by cells, to the eventual fall in blood sugar once more. Conversely, foods that break down very quickly result in a rapid process, with a sudden spike in blood sugar, and a subsequent crash down again after cell uptake. While the gradual rises and falls may go largely unnoticed by us, those spikes and crashes can cause a palpable drop in energy and a lack of focus. It's why we sometimes feel like a nap after a big meal. Eating foods and drinks that promote a gradual effect on sugar in the bloodstream, rather than a rapid response, can help to better regulate our energy and concentration levels, leaving us more refreshed and replenished.

We can work out which carbohydrate-rich foods may be better at balancing our sugar levels by how they rate on something called the glycaemic index (GI). A food's rating indicates how quickly it affects blood sugar levels. You may have heard of low or high GI foods. High GI foods break down rapidly in the gut and consequently lead to a fast rise in blood sugar, whereas low GI foods have the opposite effect – a slow rise in blood sugar due to a gradual breakdown in the gut. Examples of high GI foods are sugary foods, white bread, white rice, potatoes, and soft drinks, and low GI foods include porridge oats, various fruits and vegetables, and lentils and other pulses. However, the GI rating is not always an indicator of whether foods are healthy or not, as how they are cooked can affect how they digest in the body. For instance, if you cook carbohydrates in fat, this can lower the GI of the original food; so, crisps technically have a lower GI rating than potatoes, but we know they're not necessarily healthier! Considering the GI rating of the foods you consume may help to better regulate your energy, but don't confuse this with an optimally healthy and nutritious diet. Many low GI foods are healthy options, but we just need to use some common sense to make the right choices.

High-fat foods also have a marked impact on the brain's reward system. They can alter gene expression, giving us an increased preference for tasty foods that are often calorie-rich. In addition, long-term, regular exposure to high-fat foods appears to dampen down the pleasure response in the brain, which means you then need to eat more of them to achieve the rewarding sensation.

❷ keep your gut happy and it may keep you happy too

You may have good intentions by swapping a sugary product for one with artificial sweetener but, despite claims to the contrary, research has found that some sweeteners can also raise blood sugar. Artificial sweeteners (which are hundreds of times sweeter than sugar but contain very low or no calories) were previously thought to be inert (i.e. to have little effect) in the body and to merely pass straight through after providing a sweet taste in the mouth. However, scientists are now finding that these quiet sugar substitutes are more active than assumed. One study, from Johns Hopkins University, found that two particular sweeteners, saccharin and sucralose, induced blood sugar spikes after eating – something they were not thought to be able to do. Further research by the team indicated that the sweeteners were not causing this increase directly but were affecting the body's ability to manage its glucose levels via changes to the gut microbiome. Other studies have also found that artificial sweeteners seem to be interfering with the microbes in our gut, possibly influencing our health as a result.

There has been much talk of our gut microbiome lately. We're coming to appreciate that the vast colony of microbes in our intestines is more than just a tool for digestion. The gut microbiome is comprised of trillions of microorganisms, mostly bacteria but also viruses, fungi, and other microbes, and all their associated genetic material, and plays a key role in our health. Beyond digestion, the gut microbiome is involved in the immune system, metabolic processes, cognitive functioning, and mood. The balance of microbes influences how we feel but knowing what the right balance should be is the challenge. We're born with a certain microbiomic landscape to help us digest food and stay healthy but, as we grow, we're exposed to many factors that influence how the microbiome changes and evolves, from our genetics and the foods we eat to the illnesses we have and medicines to treat them to the stresses we experience. Our microbiome can change in response to these, resulting in health impacts.

There's much we still don't know about what the gut microbiome is doing and how. It's a tricky bit of the body to study, in part due to its location and due to its ever-changing nature. It also differs between people – everyone's microbiome is unique to them – which means trying to work out what's 'normal' or 'unhealthy' isn't easy. We're recognising that we need to maintain a contented, balanced gut microbiome for good health, but we don't yet know what that looks like or how to achieve it. We do know that certain foods, such as ultra-processed foods,[12] and medicines, like antibiotics, have a significant disruptive effect on microbes, like killing off species and altering the microbiome's activity. A particular link has been made between low diversity of microbial species

---

12   Ultra-processed foods are those that have been created from multiple processes and often contain ingredients that you wouldn't use at home. Examples include ice cream, sausages, biscuits, breakfast cereals, fast foods, fruit yoghurts, and fizzy drinks.

and elevated inflammation in the body, as well as an increased risk of metabolic disorders.

So, why am I telling you all this? Firstly, it's because if we want to feel physically well and rested within ourselves, we need to try to keep the bugs in our gut happy. But to do this, there are only some things we can control. We can't do anything about our genetics, or many environmental stressors or illnesses, but we can try to eat a healthy, varied diet and avoid ultra-processed foods, as well as limit our use of antibiotics where possible. Secondly, there is increasing research into the link between our gut microbiome and our moods and emotions. So, if we find the right balance for our own gut, it also may help our mental wellbeing.

Communication between the gut and the central nervous system was recognised by physiologists and psychologists in the 19th century but it is only in the last two decades that gut-brain axis research has emerged as a field. Understanding how the gut-brain axis contributes to health is key to explaining its role in ill-health. The gut microbiome could eventually play an important role in the development of precision medicine, with future treatments tailored to an individual's gut environment, and research has the potential to translate into dietary interventions to improve mental health, but there's a long way to go before then. Where we are now is at the level of appreciating that the gut and the brain are linked, so paying attention to what we consume could help us to influence our own wellbeing. To feel truly rested, we need our body to be aligned with our mind, and what we put into it is a critical part of this.

One brain chemical that it might be possible to influence via diet is serotonin. Serotonin is a neurotransmitter that sends messages between nerve cells and plays an essential role in many functions in the body, including helping regulate our mood and sleep. At normal levels, serotonin can help you feel calm, focused, and on an emotional even keel. Low levels of serotonin are associated with depression. Despite its influence on the brain, most serotonin is found in the gut and is important in regulating digestion and appetite. However, it's unlikely that this gut-secreted serotonin influences the brain directly as it can't cross the blood-brain barrier. While it's important to maintain healthy serotonin levels, this isn't straightforward. Low levels might be due to not producing enough serotonin or the body using it inefficiently, for example. Although there's typically a limit to how far we can proactively influence our body's chemical balance, there are some natural ways that are thought to enhance serotonin levels.

Serotonin is made from an amino acid called tryptophan, which we can only get from our diet. Eating foods containing tryptophan, such as chicken, tuna, cereals, soya beans, and bananas, is a good start but the metabolic processes involved in producing serotonin are multifaceted and also require complex carbohydrates, along with other factors, in the mix. Unlike gut serotonin, tryptophan can cross the blood-brain barrier. Some research has found that, at the right amount, increasing tryptophan in the diet can help improve mood, particularly in people who are vulnerable to low mood, as well as improve attention and memory. There are supplements on the market that aim to increase serotonin levels but there's a fine line in how much we need. Too much tryptophan is unhelpful and may have negative health effects, including impaired cognition, and too much serotonin can lead to serious health consequences, ranging from the unpleasant to life-threatening. Luckily, there are other ways to boost serotonin at a healthy dose, including getting enough sunlight every day, taking regular aerobic exercise, and having a massage, so it's worth sticking to diet and lifestyle measures that might help, without overdoing it.

'Gut-friendly' foods and drinks are becoming ever more popular. Marketeers have got wise to the importance of our guts for health and noticed the potential to drive consumer interest in products that help gain some control over this area of wellbeing. A huge range of products claim to aid our microbes and the environment in which they live, such as those containing probiotics or prebiotics. In case you need a reminder, probiotics are living bacteria and yeasts (often added to yoghurt drinks, for example), and can be found naturally in foods like kefir, yoghurt, and fermented cabbage (e.g. kimchi, sauerkraut). They are thought to help restore the natural balance of microorganisms when it has been disrupted in some way, perhaps through illness or use of antibiotics. Prebiotics are non-digestible plant fibres that nourish the friendly bacteria in your gut (these are found in many plants and vegetables). Opting for a quick fix via a handy product with *added* pro- or prebiotics might be appealing but, unfortunately, the evidence supporting them is still rather limited. Added probiotics haven't yet been convincingly proven to assist the gut microbiome in healthy people in any meaningful way. While some evidence suggests that added probiotics might help people with health issues, there's little science to support taking them to prevent ill-health. It's undeniable that we need to look after our tummies, but we can get a lot of probiotics and prebiotics from a good diet that includes plenty of plant fibres and some fermented foods, rather than relying on products with added pro- or prebiotics.

know your body

I appreciate that many of you won't really want to hear this, but the best advice is still to eat a healthy, balanced diet full of whole foods and fruit and vegetables, covering all the essential food groups. However, it isn't always possible for everyone. Millions have food allergies[13] or intolerances[14] that may result in their diets being restricted in some way. In addition, a large proportion of the global population is vegetarian or vegan, many people avoid particular foods due to their religion or beliefs, some have medical restrictions in what they can eat, some people are on weight loss diets, some people are fussy eaters, and others simply don't have access to a varied diet. In other words, it's not always easy to get all the nutrients your gut microbiome needs to keep it in optimal condition.

To achieve nutritional rest, we need to understand our individual needs. You might find yourself in one or more of the above categories and might have some idea of what's lacking in your diet or what sorts of foods make you feel less than your best, or you might just be aware

13   With an allergy, the immune system is triggered by a food or ingredient that it sees as a threat, resulting in symptoms including rash, sneezing, vomiting, wheezing, difficulty breathing, and swelling of lips, tongue, or throat, and can be life-threatening.
14   With an intolerance, the body has difficulty digesting certain foods or ingredients, causing symptoms such as cramps, diarrhoea, constipation, bloating, skin problems, and headaches.

that when you eat too much of something it makes you feel a bit rubbish even though it's not an intolerance, as such. For example, if we have a few days of eating fast food or rich meals we can feel a bit sluggish and uncomfortable. The body and brain are letting us know that they're a bit fed up of dealing with so much fat, sugar, and other heavy nutrients and would really rather have some fresh, nutritious foods and water. Anything could be a trigger for heartburn or an irritable digestive tract, for instance, whether that's spicy foods, fizzy drinks, or even particular fruits or vegetables. And what's happening in the gut is relaying messages to the brain, affecting how we feel mentally too. The key is to work out what your body needs and what to avoid so that you feel at your peak. We all have diverse requirements, so really listening to your gut, as well as understanding what a balanced, varied diet looks like, and eating accordingly, will go a long way towards helping you feel nutritionally rested.

I have no intention of encouraging anyone to limit their diets but I recommend tuning in to how different aspects of your diet make you feel. The person who knows their body best is you. It's worth reflecting on what you consume and trying to identify any elements of your diet that make you feel out of sorts, as well as those that help you feel energised, calm, or glowing. Armed with this knowledge, you can choose to better control what you eat to aid your wellbeing.

@ minimise stimulating foods and drinks

Many rely on caffeine to get the day started. It stimulates the central nervous system, increasing alertness. Going for a coffee is one of life's little joys. It gets us going when we need a lift, but too much can leave us agitated. There are tons of books, blogs, and articles about caffeine and its links with health, but I'll leave you to explore that topic by yourself. Here, we're interested in how it might impact our state of rest. On the upside, it can boost our mood and give us the kick we need to feel more energetic and focused, which may aid our mental exhaustion by increasing our concentration and motivation. But it can also negatively affect our sleep. Caffeine lasts longer in our system than we often realise, remaining for up to 10 to 12 hours, which is why health professionals often advise their sleep-deprived patients to avoid consuming caffeine after midday or earlier. This is to allow time for much of it to work its way out of the body and to give you a better chance to settle your physiology down for a calmer passage to sleep.

I'm not suggesting giving up your regular beverages altogether. After all, a hot cup of tea can be a go-to for instant comfort. In fact, tea contains a pretty unique amino acid called L-theanine that scientists have linked with relaxation. It appears to facilitate the generation of alpha brain waves and the neurotransmitter, γ-aminobutyric acid, which are associated with a relaxed but alert mental state. This is one reason why drinking tea can feel calming, despite its caffeine content, unlike coffee. It's still a good idea to switch to decaf later in the day though.

Consuming sugar in the evenings can also interfere with sleep as it also acts as a stimulant due to its effect on dopamine, so avoiding sugary foods in the hours before bed is recommended. Another substance that sparks a surge of dopamine is alcohol. In low doses, or as we start drinking it, a dopamine increase is stimulated, heart rate increases, and we may feel energised and even aggressive. Strangely, however, alcohol is also a sedative, meaning it also slows our system down. After a while, once we have more alcohol in our system, blood pressure goes down, along with heart rate, and the brain's functioning slows. This can help us feel relaxed at first, then drowsy, reducing our reaction times and physical coordination as well as mental acuity. Larger doses can suppress dopamine levels, making us feel sad or low, and eventually we fall asleep. For people who have trouble dropping off, it might be appealing to have a nightcap to facilitate the process but while alcohol can help us fall asleep initially, we often end up with a poor night's sleep. Alcohol interacts with several neurotransmitters that are important in the regulation of sleep and disrupts our natural sleep cycles (see habit 3 for details on sleep cycles), resulting in an imbalance between deep sleep and REM sleep, reducing sleep quality, as well as frequent awakenings. Dehydration from drinking alcohol leading to thirst can also result in premature awakening, as can having a full bladder from heavy drinking. Simply put, we just don't get the sleep we need.

After a poor night's sleep, some people can end up in a cycle of using caffeine, sugar, and other stimulatory substances to stay awake in the day and then alcohol later on to help drop off at night. Over time, this just exacerbates matters, potentially impacting physical, mental, and emotional wellbeing, and it becomes more difficult to truly achieve satisfying rest. If you do drink alcohol, finding the right balance for your body and circumstances is important to avoid negative effects and a knock-on impact on your rest. Because alcohol can hang around in the body for some time, it's recommended that we avoid drinking it too near bedtime.

If you're looking for a pleasure hit, there are healthier ways to boost your dopamine levels. Dopamine is made from the amino acid tyrosine, and there's some evidence that having a diet full of foods containing tyrosine may help to improve memory and mental performance. There's even such a thing as the 'dopamine diet', involving healthy foods thought to boost dopamine to enhance mood as well as lose weight. Tyrosine-rich foods include dairy items, chicken and turkey, Omega-3-rich fish, eggs, fruit and vegetables, and nuts and seeds. So, a healthy diet, including tyrosine, might help, but so can exercising, meditating, getting good sleep, and having a massage. All of which have been found to increase dopamine levels.

ⓔ try calming foods and drinks

I've talked a lot about foods and drinks that we might want to limit in our diet to avoid negative effects on our rest levels, as well as maintaining a healthy intake to avoid peaks and dips in energy, but there are also foodstuffs that might actively help us to feel calmer and more rested.

Firstly, and most importantly, drinking plenty of water is an essential part of wellbeing. It keeps all our cells hydrated, facilitates digestion, regulates our temperature, and enables a healthy flow of nutrients around the body. Too little water can give us a headache but also impair our concentration and energy levels, as well as leave us feeling sleepy. To feel optimally rested and refreshed, most of us would certainly benefit from drinking more water.

Some research suggests that foods containing magnesium, like leafy greens, nuts, and wholemeal bread, may help us to feel calmer. Dietary fibre has been found to reduce inflammation in the body and brain, and increasing its consumption, through fruit and vegetables, whole grains, and legumes, for instance, might dampen down the inflammatory response. Foods containing Omega-3 may also help to lower anxiety, through mechanisms involved in inflammation, so upping your intake might help to relax your mind. Iron deficiency has a significant effect on our levels of fatigue, and can lead to headaches, poor immune function and memory, amongst other health issues. Iron deficiency anaemia is also linked with mental ill-health, such as anxiety and depression. So, ensuring we have enough iron in our diet is critical. Other nutrient deficiencies associated with low mood, anxiety, or depression include folic acid, vitamin B12, and vitamin D. Increasing the amount of foods containing these nutrients in your diet might improve your rest levels. However, much of the research into the effects of nutrients on mental wellbeing is patchy and hasn't been extensively tested in real-world diets so none of this is for certain. There are links in the evidence and a few trials of component ingredients but little that indicates direct cause and effect, so we need to be cautious in how we interpret the science. On the positive side, you'll notice that the foods thought to be helpful are pretty healthy options in moderation and increasing the intake of these, for the average person, is likely to be beneficial to health in a host of ways.

What is not being suggested here is taking additional supplements. The problem with many supplements is that they often haven't been proven to be beneficial and there's a chance of negative health effects from taking too high a dose. For healthy people,[15] it's always better to improve the diet through food and drink rather than supplements, where possible. An exception might be vitamin D, for which most of us cannot get our recommended daily amount through diet and sunlight alone.

You may find that there are already foods or drinks that are calming for you, whether that's a herbal tisane, warm milk or cocoa, or even mashed potato. Whatever it may be, if you've found a harmless way to improve your rest levels and wellbeing (without being too unhealthy), then stick with it!

---

15   This excludes those who have been advised by a health professional to take particular supplements due to their specific health circumstances.

☺ challenge how you think about food and drink

How we think about food and drink can also affect how we feel after consuming them. They might make us feel happy (perhaps a delicious dessert), guilty (possibly a cheeky mid-week takeaway), or virtuous (maybe a simple salad) and this can influence our mood. You may even think of certain foods as rewards and others as punishments. Let's say you've been on a bit of a health kick lately but give in to a calorie-laden couple of days. You might now feel that you've let yourself down, in some way, or your motivation to keep being healthy is waning, and this could result in low mood. Studies have found links between eating junk food, for instance, and negative moods, as well as a greater likelihood of showing signs of depression. The relationship between fast food and mood is complex but evidence suggests that fast food can alter our state of mind due to negative connotations we may place on the foods.

Foods are often closely associated with our emotions (the phrases 'emotional eating' and 'comfort food', for example, are familiar to most of us) and can make a big difference in our feelings of rest. Eating well, consistently, can help but so can altering our perceptions of foods. Considering specific food to be 'comforting', associating contentment or wellbeing with its consumption, may be unhelpful as it makes us want it more yet it may not be particularly nutritious. Equally, considering other foods as 'bad' in some way may lead to feelings of guilt or disappointment if we eat them. Having a healthy attitude towards foods and drinks, by trying not to place too much weight on their emotional connections, can lead us to be more positive and help us take control of how they make us feel. It's all about knowing what we need, versus what we want, and finding a healthy balance. You only have one body, so treat it well and you'll notice the benefits.

We often use foods and drinks to prop up, or control, our moods, pleasure levels, and energy in the short term, but we must remember that in the longer term these habits may come back to bite us. From energy slumps to sluggish digestion, mental fogginess, memory impairment, and poor sleep, let alone the impact on our physical health, in failing to dedicate attention to what we consume we may miss a substantial opportunity to improve our health and contribute to our overall rest. Good nutritional habits can help address issues with sleep, energy, concentration, mental and physical performance, immune functioning, mood, and more. Reframing our attitude towards different foods may also enable us to manage our emotions and mental state better.

As we discover more about the relationship between how we eat and how we think, tools to help us understand, and adapt, our own eating patterns are being developed. Apps are using psychology to help us lose weight, for example, by changing our mindsets around food to form healthier habits. It's thought that by changing how we think about food, we can modify how we eat and transform our way of life through long-term behaviour changes.

● try mindful eating

An aspect of mindfulness that is frequently alluded to is eating in a mindful way, using all your senses to experience food to its fullest. This involves focusing solely on your food, or drink, as you consume it, from looking at it and noticing its form, colour and texture, for instance, to tasting it and detecting any sensations, foretastes and aftertastes, and any sounds, then perceiving how you feel after you swallow it. You take your time. You're literally savouring the moment. It's not suggested that you do this for every food and every meal – that would be going too far for most people. There are simply not enough hours in the day to live like that! But the idea is that by trying it now and again, your mind is brought entirely back to the present moment, and by concentrating on what you're consuming, you intensify the experience and heighten your awareness. You take time to really enjoy your meal.

I've tried mindful eating a few times, as have friends of mine, and we've all noticed things we hadn't before and found flavours and textures to be amplified – foods to be sweeter or crunchier than we had realised, for example. By doing this, you might find that you eat less of a food to get a sugar hit as the sweetness seems more apparent, or you enjoy certain foods more because you notice the complexity of their aromas or flavours, or your mood alters as a result. It might even change your eating behaviours or food choices. The great thing about mindful eating is that it is done without judgement. So, not worrying about what you're eating but just bringing your attention to any observations about it. It might help you to better notice physical hunger and when you're full, or when you're emotionally eating. Rather than being about the foods themselves, a mindful meal is more about the process of eating. Better appreciating how we eat, as well as what we eat, might help uncover why we eat what we do. It helps us to better know ourselves, and that can be the first step if we feel we need to change our eating habits to improve our nutritional rest.

what did *the panel* try?

1. Eat and drink healthily for at least three consecutive days (including no caffeine after midday, no alcohol, no fast food or takeaways). Only a few volunteers had a go at this as it required more effort than many other of the activities but, for those who tried it, it made a noticeable improvement to their energy levels and general sense of wellbeing.

This is pretty much the way I eat anyway, and certainly during the weekdays. On weekends, I might have one takeaway or eat out and also have more unhealthy treats or desserts, but Monday to Friday I keep it pretty healthy. I don't really drink much alcohol, even on the weekend either. I feel so much better in myself and have more energy when I eat and drink like this, so have stuck to it for a long time now – **Caroline.**

2. Eat in silence and mindfully experience a meal. Several Panel members had a go and noticed that it heightened their awareness of what they were eating and the moment itself.

I've always known that I wolf down my food, and making the effort to eat mindfully had the result of slowing me down. I realised from this activity that meals have become purely about getting energy into me, despite the fact that eating dinner with my partner is one of my favourite things in life. Whilst I get excited about what is cooking, it ends up being quick gratification rather than savoured. I did feel cheery when I noticed all the different colours on my plate and felt grateful at times for having the lovely food that I do. But I often found myself ruminating. I had to bring my mind back many times. To make it more feasible, I made an effort to just look at the food on my plate before I got lost in the eating. I just noticed the moment before tucking in. This felt really nice and is a good start. I'm going to do this more and just take in the whole moment, not just the food but also the fact that I'm with my partner, it's the end of the working day and we're starting to relax. Making it more about the situation, I think, works for me – **Louise.**

I felt driven to have something more interesting to eat, so rifled through the fridge and prepared a satisfying meal. I really noticed what I was doing in the moment and how I experienced the food – including how it looked, its texture, and taste. I found it difficult to fit in, as other things were going on, but I did enjoy reading one lunchtime, which felt like a break. This activity seems like it's good, in terms of not over-eating and taking a mindful time-out, but one of the pleasures, to my mind, of eating alone, is that one can read and relax in other ways – **Duncan.**

habit 5 :

tune in to your senses
for *sensory* rest

"Unnecessary noise is the most cruel absence of care
that can be inflicted on the sick or the well"
Florence Nightingale

Do you ever feel like you're just going through the motions, hardly distinguishing one day from the next? Time is flying by without you really noticing it, leaving you feeling hollow or a bit blank. What's all that effort for if you can't also enjoy the life you've worked so hard to build? And do you find that the world around you is taking its toll on your energy, with all the demands it places on your attention? We can become so preoccupied with our day-to-day responsibilities that we forget to take time to appreciate the here and now, and we may find ourselves so overloaded by sensory onslaught that when we want to enjoy the world, we can no longer find the head space to welcome it.

Achieving sensory rest involves taking time to listen to what our body is telling us and making positive choices to enhance our sensory experiences of life. It's about tuning in to our senses in such a way that it increases wellbeing and enhances a sense of calm. If we are to be more in control of our inner tranquillity, we need to find ways to experience, and connect with, our senses on our own terms; to dim the stressful sensory experiences that exhaust us and amplify the beneficial ones that make us feel good.

what might help?

℮ recognise the impact of noise

I couldn't agree more with Florence Nightingale when it comes to noise. It can be so intrusive and exhausting, and significantly impact well-being. Noise isn't just a nuisance – it's bad for our health. Putting aside the obvious harmful damage of loud noise on hearing, regular exposure to unwanted noise has been linked to other health harms. For example, the World Health Organization estimates that exposure to traffic noise is responsible for the loss of more than 1.5 million healthy life years per year in Western Europe alone, largely due to annoyance and interrupted sleep, as well as cognitive impairment in children and ischaemic heart disease. Traffic noise has also been related to high numbers of strokes, heart attacks, heart disease, high blood pressure, as well as depression and anxiety. Assault from prolonged environmental noise results in sleep disturbance and disruption to daily activities, increasing mental stress, and raises stress hormone levels, leading to inflammation and a cascade of other negative physiological reactions throughout the body, which then accelerate the development of health problems.

It's not just traffic noise that's getting us down. Noisy neighbours are frequently cited as a cause of stress in health surveys. Danish Health and Morbidity Surveys, from 2000 to 2017, revealed neighbour noise annoyance in multi-storey housing was higher than annoyance from traffic noise. In 2017, 36% of respondents reported having been bothered by neighbour noise and 22% by traffic noise, and neighbour noise annoyance was strongly associated with health outcomes, such as fatigue and sleeping problems.

I once wrote an article about noise in hospitals being an impediment to patient recovery, as it's such a widespread issue. Noise generated by machines and staff has been found in several studies to disturb patient sleep and escalate irritation and fatigue, and has also been linked with high blood pressure, increased pain sensitivity, and poor mental health. The very place that we rely on to provide a suitable environment to support recovery is literally making people more ill, simply through noise pollution.

When we consider the core purpose of hearing – to alert and warn – it's no surprise that sound is so closely connected with our emotional reactions. Noise triggers a response in the brain's amygdala, which is a centre for both our emotions and stress, also regulating our fight or flight response. But the type of noise, or sound, matters as our brains respond accordingly. After all, we know that while some sounds are infuriating or frightening, others might induce joy, serenity, or laughter. A US team of researchers inserted electrodes inside the brains of fifteen people who were due to have brain surgery to analyse the responses of their neurons to different sounds. What they found was fascinating – that there are groups of neurons that respond specifically to the sound of singing. Although located in a similar part of the brain, other groups of neurons selectively responded to speech or instrumental music. What the study revealed was that our brains engage with sound in a very nuanced way, being able to selectively detect and react to diverse sounds, which may help to explain why different sounds lead to very different physical and psychological effects in us. It's been suggested that specialised groups of neurons have evolved to enable us to quickly hone in on particular sounds in noisy environments, allowing the brain to ignore background noise.

Sometimes we can feel that everything is too noisy. Every little sound grates on us – the sound of someone chewing, the incessant tinny beats coming from a fellow commuter's headphones, the siren of an ambulance going by, the sound of a hand dryer or vacuum cleaner. It's tempting to dismiss this as being grumpy or intolerant, but noise sensitivity is a real thing. 'Hyperacusis' is when everyday sounds seem much louder than they should, and it affects a significant number of people. Such sensitivity can affect your mood and wellbeing, with a knock-on effect on relationships, work, or school, for instance. It might be something you've always had, develop over time, or come on quite suddenly, perhaps due to a health condition, such as migraine, Lyme disease, or a head injury, but often there's no clear cause.

Emerging from extended periods of lockdowns during the Covid-19 pandemic, many people reported that they now felt very sensitive towards noise. Being in our own homes and away from the world at large meant that our ears and brains got used to being in much quieter environments, with far fewer sounds to process. At the same time, the world became a much quieter place as traffic stopped, planes were grounded, businesses closed, and people left the streets. We could now hear nature as it really is – bird song, buzzing insects, leaves rustling in the breeze. This in itself was a joyous discovery and great for our sense of calm, but getting back to normal meant we had to then desensitise ourselves to the noises of everyday human life. For those who suffer chronically from hyperacusis, help to reduce anxiety from noise includes sound therapy to get used to sounds at normal levels again, and cognitive behavioural therapy to alter how to think about sounds. The aim is to increase the ability to tolerate sounds, and accept things the way they are, to aid your own serenity.

☙ embrace the sound of silence

Sometimes, we need to get away from noise altogether. Norwegian adventurer Erling Kagge, in his escapist book *Silence: In the Age of Noise*, explores the power of silence and how he experienced it on his solo adventures. It has to be one of the quietest and most relaxing books I've ever read, and I thoroughly recommend it, but many of his reflections come from physically being in some of the most remote places on Earth (e.g. the South Pole). On the one hand, this gives him great insight into true silence and its effects, but on the other hand, it's not something many of us can truly relate to as finding such silence is a luxury we can only dream of.

It can be difficult, and sometimes impossible, to locate a quiet space. The noise of urbanisation (vehicles, construction sites, alarms, people, music) and technology surrounding us (phones, computers, smart-everything) intrudes into every moment of the day and can be hard to avoid. Beeping, buzzing, hammering, humming, vibrating, shouting, barking, revving, thumping, squealing... the sounds keep coming. Much of the time we fade them out, because if we actively listened to them all it would take up far too much energy and brain attention. With a bit of practice, humans can get quite good at focusing on tasks while exposed to background noise and this is an essential skill to be able to function within society. For example, this helps when working in an open-plan office, concentrating in the classroom while other students are at play outside the window, or getting drinks orders correct in a bar with loud music playing and multiple customers demanding attention. Yet, just because we can juggle these sensory burdens, it doesn't mean that we're okay with the status quo. Most of us probably recognise that we need some peace and quiet now and then, but science tells us it's actually really important for us to regularly seek silence out.

Exposure to welcome[16] silence has a profound effect on both our physical and mental health. It's been found to lower breathing rate, blood pressure, heart rate, and cortisol levels. Researchers from the University of Edinburgh found that 15 minutes of silent relaxation was as effective in reducing post-operative pain as listening to calming music. Even just short periods of silence, of just a few minutes, can improve mood, promote calm, and alter our perception of time. Silence can adjust our awareness to the present moment and is often intertwined with mindfulness. As we seek to bring our attention to the present and connect with our environment, exposure to silence during this time may heighten our experience of sound, as well as our other senses, as the brain is freed up to tune in to what we're feeling and experiencing (see habit 1 for more on this). Some researchers propose that it is not silence itself that is good for us but the absence of noise, which can be harmful – a bit like saying that not smoking or not jumping in front of fast-moving vehicles is good for us. In other words, it is the absence of the harmful factor that is beneficial. This is something that is challenging to prove because we don't live in a vacuum. It's also possible that silence is good for us independently of noise being bad for us. Whatever the reason, dimming noise pollution from our lives and purposefully seeking suitable periods of quiet can only be helpful.

Despite gaining his epiphany from a place of ultimate isolation and silence, Erling recognises that such an environment is not achievable, not only for most of us but also for him in his day-to-day life back in Oslo. As such, we all need to create silence for ourselves. He believes we can all find ways to dim the noise of modern life and find moments of silence.

16   I say this because it's not always welcome, and may stem from isolation, loneliness, or trauma for instance, and can then have negative effects.

What he's referring to is not about attaining the absence of sound but the ability to tone down background noise and distractions. It's about being in your own thoughts, in the moment. Just like when you're doing a hobby you love – maybe reading, playing a guitar, cooking – you're focused on what you're doing and not thinking about worries or planning; you're just doing and being. That's how I feel when I swim lengths in a pool. Once I get into the right pace and rhythm, my mind switches off and I lose track of time and awareness of the outside world. For that space of time, I'm in my own realm, experiencing inner silence (according to Erling's point of view). He suggests that you follow your own path and discover yourself along the way. This might seem a bit wishy-washy to some, but is akin to the process of daydreaming, meditating, and simply relaxing.

As Erling illustrates, silence doesn't just have to come from spending time alone in a room being still. We can achieve silence when in nature and even in a group, but it may be easier to find silence and space to think when we're alone. However, despite the potential benefits, time alone and in quiet is often considered a negative experience. Being in silence, sitting with your own thoughts, can be uncomfortable for some people and might spark a sense of boredom or unease. It can create space for unhelpful repetitive, insular, and circular thinking – in other words, it can leave you wallowing in negative thoughts. It can even feel like torture if you're waiting for someone to speak, or to visit, or if you have so many worries that the last thing you want to do is to be left alone with them in your head. Studies have even shown that some people would rather give themselves an electric shock than sit by themselves in silence for 15 minutes, for instance, and loneliness is at an all-time high at great detriment to health. But being comfortable in our own company, engaged in the moment and our surroundings, including finding silence, is a skill that can reap its own rewards. Taking time to find short periods of silence a few times a week could improve your self-awareness and levels of contentment and induce calm in turn. Intentional, undistracted solitude can help to slow us down, disrupt automatic negative behaviours (like scrolling through social media on autopilot), and promote thinking from a different perspective. We should strive to alleviate the impulse to constantly fill the gaps in our daily activities with distractions, like phones, TV, gadgets, or unnecessary chattering. Give yourself time and space to be alone (without external distractions) or to be quiet in the company of others. Being in nature is a good way to start, as it is a pleasant and relaxing place to be and has even been shown to reduce the urge towards boredom.

Society is also trying to find ways to help. From quiet coaches on the train and better soundproofing of hotels and workplaces to public policies limiting amplified sound levels, efforts are being made to improve noise levels we're all regularly exposed to. Technology has also jumped onto the trend – after all, if there's a buck to be made…! Huge numbers of apps and gadgets claim to help in the battle against noise. From noise-cancelling headphones and white noise generators[17] to online guided meditations with deliberate periods of silence, there are plenty of options to try. Many of these focus on the sensory aspects of rest, such as sound, but there's a relative dearth of evidence to back them up. Marketing hype combined with the pressing desire for better wellbeing is all it takes for some – whether they work for you or not is for you to discover.

℮ try a sound bath

Sound bathing is a form of meditation that aims to promote relaxation and tap into emotions through immersion in sound and vibration (akin to immersion in a warm bath, although no actual bath is involved). Participants lie comfortably on a mat, with a pillow and blanket, while someone strikes gongs and other percussive instruments at intervals during a session of anywhere between around 30 minutes and two hours. It's said to be more a sensation than a sound, that flows into the brain, and those who've tried sound baths describe the experience as like ripples of energy passing through the body, going into a trance-like state, or achieving a sense of calm, with an increase in the feeling of wellness or inner peace. It is also claimed to help release emotions, with some participants crying or laughing during a session.

This was something I was keen to experience and like nothing I'd ever done before. There are thousands of online options, so I started there, in

---

17   White noise is a constant background noise, containing all frequencies, that drowns out other sounds.

the comfort of my own home. I experienced a tingling sensation right through my body down to my toes at one point (intriguing!), so I tried out different sound bathing sessions, by different practitioners, at different lengths. Overall, the experience was relaxing and enjoyable but it felt like something was missing – the physical sensation of the sound vibrations. Next step: an in-person session.

I was lucky enough to try an outdoor and an indoor sound bath, both in beautiful woodland settings. The outdoor session took place on a gorgeous, balmy summer's evening in a reconstructed Anglo-Saxon village.[18] It was very pleasant but I was distracted by the soft breeze blowing my hair across my face and an ant crawling over my bare feet. The indoor sound bath was a world apart. It was cosy and intimate, with seven participants armed with a huge array of items to get comfy.[19] The small room was lit with fairy lights, and logs were piled in a fireplace that gave off a Nordic lodge vibe. After a brief meditation to settle, the sounds began. A range of instruments were used in succession, creating overlapping layers of continuous sound that built in intensity, causing vibrations to reverberate off the walls and back across the room. The effect was powerful – the sounds went right through me, my heart started to race as the volume escalated, and my mind had a million racing thoughts,[20] then I suddenly fell asleep and drifted in and out of consciousness until the end of the session as the sounds softened and slowed. I left feeling incredibly relaxed and sleepy. My mind was quiet and content, but the most surprising effect was how peaceful my body felt, as though I had had a massage. About an hour after the session I felt even more rested, which was something another attendee mentioned that he often

---

18   My lasting memory of this place was from my school days when we visited, dressed as Saxons, and I was hit round the head by another kid whose parent thought it was a good idea to make them a sword out of a fencepost. I only had a flimsy cardboard one to defend myself. Good times!

19   Impressively, one woman brought a hefty memory-foam mattress topper, and another wore a chunky, woollen polo neck jumper and thick woollen socks while lying under a duvet. Reader: please note that it was a hot August evening!

20   This slightly reminded me of 1960s spy movies in which layers of weird sounds are used as a weapon.

experienced. I wanted to understand more about sound baths so I chatted to the session leader, Nicky Dorrington [see box].

I was interested to know how Nicky got into sound bathing and why.

I knew I needed to change my way of life... I was busy commuting, super stressed and time poor. So, I first turned to very physical yoga and simultaneously realised the benefits of meditation. However, I have such a chatty mind and found traditional seated meditations challenging. I could be meditating, and my mind would be making shopping lists three weeks in advance! A lot of people find it difficult to sit or lie and quiet the mind to simply be with their thoughts, so I started to explore the concept of sound bath meditations. I'd read that it can be deeply relaxing and might help if you struggle to meditate, so I thought I needed to try it and went to a session. From the first time, I was completely blown away by how timeless the hour felt and how the sounds really did cut through my mental chatter. With no effort, the sounds had transported me to a state of deep relaxation and meditation, and I was intrigued. From then on, I used sound meditation as a tool to manage stress, along with yoga nidra and physical yoga practices, and these massively helped me cope with the demands of my life and my job.

Looking back, if you had told me I would be teaching meditation and running sound baths I would probably have just laughed because my job was so corporate and a world apart. But then I had a baby, with a difficult birth and breastfeeding trauma resulting in PTSD. It was honestly brutal and I struggled with my mental health for a while. I felt pretty broken but through this adversity came an opportunity to reassess my life and think about what I truly wanted to do. Alongside traditional talking therapy, I turned to nature, sound, and meditation to see me on the road to recovery and they were so pivotal to my recovery that I knew I had to share these practices with others. I trained as a sound therapy practitioner, forest bathing guide, and meditation teacher, specialising in yoga nidra and bringing nature into everything that I do.

Life experiences often take us on a different path from where we started out and I was keen to understand what sort of people are drawn to sound bathing.

Most sound bath attendees have already discovered the power of deep rest. Newcomers are typically at the 'things have got to change' junction in their lives, having perhaps realised that they've been putting themselves last on their to-do list for too long. When attendees start to intentionally rest, I often hear that they left the session feeling better equipped to cope with the demands of life. This can be transformative. It's not uncommon for guests to feel emotional in the sessions – having that space to breathe and just be, accepting themselves as they are with no expectations of them in that particular time, can feel like a release and is really powerful. Sound, in particular, can be evocative. When we quiet the conscious mind, thoughts and emotions from our subconscious mind have space to surface. Commonly, these are all those emotions we've held down for a long time. This emotional release can be part of the journey to self-acceptance and rest.

While I'm definitely a convert and wish we could all regularly immerse ourselves in sound baths, this clearly isn't realistic, so I asked Nicky to suggest tips for calming us in our regular lives.

Always come back to your senses and your breath. These are with you wherever you are and if, at any point, life is getting a bit too much, tune in to your senses: 'What can I see, hear, feel, taste, and smell?' This helps to bring us into the present moment. Breath is really powerful. We're often told to take deep breaths when we need calm, but steady breaths while lengthening the exhale is a much more effective way to regulate your nervous system. Secondly, any opportunity you can, kick those shoes off and get your feet in the earth. Notice how it feels. It can, again, help you feel in the moment and more grounded. And, thirdly, get connected with nature. Pick up something from nature, whether that's a leaf, a feather, a pine cone, and really look at it.

> *Observe and marvel at its form, feeling, colour – anything you notice. These are all things you can do any time.*
>
> Just listening to Nicky talk is extremely relaxing and makes me want to immediately head off for a lie-down, with only gong sounds for company.

Sound healing[21] has been in use for thousands of years but in more recent times, scientists have been delving into what we really know about it. Some research has shown a measurable effect on improved mood, relaxation, and reductions in tension and fatigue, but the studies are fairly small, observational, and typically non-randomised – participants were likely to want the interventions to work and may have been more suggestible. However, as there don't appear to be any significant harms for a healthy, non-pregnant adult, it may be worth giving it a try to see for yourself, whether in person or online. One benefit of sound therapy is that it doesn't require you to learn a disciplined form of meditation. Mindfulness and meditation take time to learn, practise, and get the hang of, so if you can achieve similar effects more quickly, by taking part in a sound bath session, for instance, this may be preferable for many who are time poor and desire more instant results.

◉ let the music in

There's no doubt that hearing music can provoke strong reactions. From cheerful to melancholy, playful to focused, and finding solace to nostalgia, music is a powerful influence over body and mind. During the Covid-19 pandemic, people turned to music to help cope with anxiety, uncertainty, isolation, and the toil of living in confinement. From communities singing on balconies to online choirs, bands, and concerts, huge numbers were getting involved in making and sharing music, as well as consuming it, to manage their wellbeing. Researchers found that people spent more time

---

21   Sound therapy should not be confused with 'music therapy', which refers to an established psychological clinical intervention, involving the application of music, by trained professionals, to help people with specific needs.

engaging with music during the pandemic than before it, often as a way to help regulate their emotions as well as to facilitate social interaction (albeit virtual). This was definitely the case for me. While others burned their pent-up energy through a daily online physical workout, I danced around the kitchen to music – it doesn't just encourage movement to keep fit but also boosts my mood and reduces stress. When I need uplifting, I often turn to dancing and singing along to music. When I can't motivate myself to move or sing, for me that's a clear sign that I'm not feeling on par emotionally.

Music has been shown to profoundly alter the listener's internal state as it affects emotion, cognition, physical activation, and physiology. Music is a potent tool that we can employ at will to influence how we're feeling. Just think of lullabies sung to small children and how they encourage sleep (often very effectively in the tired parents too!). Numerous studies indicate that listening to music is good for our wellbeing, with beneficial effects seen in mood, relaxation, self-esteem, social connection, movement, and reduction of stress and pain, for instance. Lots of people listen to music daily. This may be passively hearing a selection of music in the background (such as from a radio) while doing other activities, or it might be deliberately chosen to match or improve the mood of the listener (like choosing angst-ridden music during the break-up of a relationship, or upbeat music when you feel like partying). We often associate particular styles of music with relaxation – certain tempos, instruments, and vocals encourage a calm atmosphere, depending on our own musical preferences. Actively choosing music suited to your mood is a simple way to encourage sensory rest. It could also help lessen the stress of any activity you may be doing at the same time.

I would argue that music is essential for the soul and the great thing is that it's universally accessible. Anyone can sing or dance or tap out a beat for free. You don't have to be good at it or know what you're doing; simply humming a tune to yourself counts! Even people with hearing impairments can feel rhythms and move to music (just look at Rose Ayling-Ellis, the 2021 celebrity winner of the UK's *Strictly Come Dancing* TV programme, as an inspirational example). Whether you're into teen pop or Beethoven's 5th Symphony, folk music or 90s dance tracks, avant-garde jazz or bhangra, let the music play and embrace its power.

As well as music, some are choosing to listen to 'binaural beats' to relax. These are tones of different frequencies played in each ear to create the perception of a new tone (or beat) within the brain (it gives the effect of hearing something that isn't really there). Small studies of brain images show some changes during listening to these beats, but there's no evidence yet that this is beneficial in any way. Any restful or meditative effects may be similar to those of simply listening to calming background music.

But we should avoid using music too often to shut out the world. There's been a big increase in the number of people who seemingly have their headphones glued on all day long, whether working, walking, or travelling, for instance. While listening to music can be relaxing and a respite from disturbances or stresses around us, it can be an escapist habit that drives us too often into a more insular place, reducing our connection with people and our environment. This leaves little room for mindful awareness, which may reduce our resilience and ability to cope with everyday happenings around us. A healthy balance is required for good wellbeing.

It's not just listening to music that can make us feel good. Singing is well known to aid our psychological wellbeing. As well as boosting mood and reducing stress, it's associated with improved confidence, self-esteem, and empowerment. It helps with memory and it produces pleasurable sensations, activating the brain's reward system so that we're prompted to do it more often. Singing has been found to produce a greater effect on increasing positive mood than just listening to music. It also helps with physical movements that are known to improve a sense of calm, including controlled breathing, posture, and relaxing muscle tension. Learning to play an instrument has also been linked with improved cognition, self-esteem, and wellbeing in studies across diverse populations, and, depending on the type of instrument, can also help with breathing, posture, and muscle tension, as well as promoting positive feelings.

☺ create a calm personal environment

Is it just me, or do you find being surrounded by mess and clutter stressful? It's a visual indication of jobs we need to do (wash the dishes, do the laundry, sort out the bills) and a distraction from things we would rather attend to. Maybe you want to sit down to a family meal but the table needs clearing first, or perhaps you long to have a quiet sit and read but dust on the bookshelves or toys all over the floor mean you can't focus. Perhaps the mess is a reminder that someone you share a home with hasn't done their chores, increasing your frustration, or a pile of old paperwork is getting in the way of fully concentrating on current work tasks. For some, having a clear home and work environment just makes them feel calmer and more in control. Research has found that clutter can create too many stimuli and overwhelm the visual cortex, making it more difficult to focus. It has also been linked with depression in people who feel it somehow represents unfinished tasks and can make people feel self-conscious and embarrassed about inviting visitors into the home for fear of being judged by their mess. The perception of lack of control, driven by not being able to manage the clutter, also leads to stress.

Mess and clutter contribute to general anxiety, frustration, and confusion. If you can't find that thing you're looking for because of all the mess, you might question where you left it and whether you really left it there. You might forget what you intended to do because you got distracted with a different task due to the clutter in your line of sight. Sometimes the mess can feel so overwhelming that you don't even know where to begin so you hardly bother tackling it at all – it's just easier to ignore and work around it. Getting on top of our belongings and the things around us can make a big difference to how we feel. Organising our stuff can help to reduce stress and mentally create room for things we would rather spend time on. Tidying and decluttering help us gain some control over our environment and not allow it to overwhelm us, as well as create a more serene space to inhabit. This isn't about minimalism, which is more a lifestyle choice, but simply about clearing out things you no longer want or need and organising your belongings in such a way that supports your wellbeing.

The most famous proponent of decluttering is Marie Kondo. She created the KonMari method, which prioritises what to keep rather than what to discard. She says if something makes you feel joy you should keep it, and if it doesn't then you get rid of it. Marie started her first tidying business as a 19-year-old university student in Tokyo. Since then, she's written nine books, had an award-winning Netflix TV show, and now trains others to become KonMari consultants and has her own online homeware shop (where you can presumably buy more things you don't really need!). Tidying is clearly big business as it's something people want and need in their lives. We're all striving to feel better and live better, and Marie's approach tempts us with a lifestyle many wish to achieve.

The KonMari decluttering method first asks you to visualise what kind of space you want, and what kind of life you want, before you can declutter effectively. You're putting in place the goal to aim for. Once you've enlightened yourself on what you really want, you can then sort all your belongings based on whether they make you happy (or 'spark joy') – this includes making you happy because they have a clear purpose, in case you're wondering whether to throw out that toilet brush! Items are sorted based on a category, rather than a location, starting with the easiest (clothes), followed by books, papers, and miscellaneous household items, and leaving sentimental items to last as they can be the most difficult to part from. By working in this way, Marie believes that you will ultimately be left with a home of things that speak to you and the way you really want to live. It purports to be a mindful and forward-looking approach that empowers you to keep control of the clutter.

While the KonMari method is admirable, it's pretty full-on, and you don't have to go through all your belongings to start to feel better and calmer. Clearing a whole home can be overwhelming and stress-inducing in itself, so, as with all organising tasks, the best way to approach it is to break it down into manageable chunks. Start by spending 15 minutes clearing just one small space at a time, such as a desk, a shelf, a window-sill, part of a kitchen worktop, a drawer, or one pile of papers or toys. A bit at a time can add up to a big difference in how you feel. When you want to relax, you can then face the direction of that clean space while undertaking a restful activity. Once you've decluttered, it's important to try and keep on top of things – try not to let jobs pile up and avoid leaving everything for a later time as you may end up back at square one.

My friend Heather hired a decluttering professional, and simply committing to action improved her mood immediately.

We live in a very old house that has been in our family for a long time. When we moved in, there were many items already there that had been passed down through generations. Many of the items we don't need, but we have to sort through everything to work out what stays and what goes. It's such a big job that has been hanging over me for six years. It felt really overwhelming, given that I also work and we have two young children. I finally recognised that I couldn't do it all by myself and I hired a professional declutterer to help. Just booking her in made me feel instantly better, as though a weight had lifted off my shoulders.

Making progress over a number of months transformed her stress levels and set her on a path to greater control over how she wants to live.

Since I've had my declutterer come every month to work on a different part of the house, I feel so much better. It makes me so happy to see our belongings sorted out properly and I feel much less anxious about having it hanging over me. Some of my friends say that I could've just done this myself without getting in a declutterer, but the reality is that it just wouldn't have happened. We've lived with the mess for years but having someone come in and help has not only made something happen, but it's made me accountable for dedicating attention to getting the clearing out done.

Funnily enough, in 2023 Marie Kondo revealed[22] that she's 'kind of given up' on having an immaculately organised home since having three kids.

---

22   Banfield-Nwachi M. *Queen of clean Marie Kondo says she has 'kind of given up' on tidying at home.* Guardian online. 30 January 2023. www.theguardian.com/lifeandstyle/2023/jan/30/queen-of-clean-marie-kondo-says-she-has-kind-of-given-up-on-tidying-at-home (accessed January 2023)

For her, the priority now is spending quality time with her family, with tidying a secondary concern. She's shifted in her approach to something a little more realistic for most of us; rather than paring back your belongings to a bare minimum, it's about evaluating how you organise and order your life and finding your own path based on what fills you with joy.

Another Japanese concept that wellbeing gurus have leapt on is *wabi-sabi*. It's based on a soulful and simple way of life, that appreciates frugality and humility, as well as simple beauty, imperfection, and the transience of things. This certainly sounds like an aspirational way to live, and one which would no doubt calm us all. It's difficult for the average person to pick up such a concept and fully embrace it, but one element that we could learn from is that a *wabi-sabi* home values decluttering. It is not a focus on minimalism, as such, but on what truly matters and is important to you and those things that evoke positive feelings. The idea is to encourage us to simplify our lives, so we give space to those things we cherish. And that is surely something to aspire to.

There's more we can do to create a calmer environment beyond tidying up. How we decorate our surroundings and the things we put in spaces can impact how we feel about, and use, them. Depending on personal tastes and needs, different colours, textures, and lighting can evoke moods more or less inducive to calm. There's a reason why day spas feature pale pastels and neutral shades, and nightclubs tend to opt for dark and bold colours. If places where you spend a lot of time feel stressful to be in, see if there are any changes you can make to their physical appearance and arrangement to make them more pleasant and relaxing. Can you paint the walls or put restful images on them? Can you reduce the clutter and introduce clean lines or softer textures? Could you alter the lighting to something less glaring or more natural?

Aromas can help too. Not only can they give a room a pleasant smell, essential oils have been found to have a wide array of beneficial effects on our physiology through gentle inhalation. These include antimicrobial, antioxidant, and anti-inflammatory effects, benefits to the respiratory system, memory, cognitive performance, and quality of sleep, as well as reduced anxiety (see habit 6 for more on essential oils). Not only can certain scents help us relax, they can also spark memories. When you smell salty air, perhaps you think of a particular beach trip and people who were there, or the smell of certain foods or even washing powders may trigger a memory of a significant time or person in your life. Some may be fond memories, but other smells can dredge up negative feelings, and offensive odours can also result in discomfort and stress. Clearing a space of unwanted smells can help defuse tension.

Strangely, feelings of anxiety and stress can also cause things to smell bad. In one study, participants were presented with a range of aromas to rate how they felt about them. The odours were neutral and inoffensive and, initially, were rated by participants as such. Then, to induce feelings of stress, they had their brains scanned while being exposed to disturbing images and text messages, and then were asked to rate the same smells again – this time most subjects changed their rating from neutral to offensive. The scans suggested that negative emotions caused by the disturbing images and messages sparked a 'cross-wiring' between two areas of the brain, the smell circuit and the emotion circuit. According to the scientists, when we're exposed to odours, normally it is just the olfactory system that is activated, but when an individual is stressed or anxious, the emotional system becomes part of the olfactory processing stream. Additionally, it seems that this cross-wiring builds up over time, as we experience more anxiety, strengthening the connection between smell and our emotions, with the result that more typically neutral smells are perceived as negative. Those negative smells then contribute to our anxiety, triggering a vicious circle of smell and stress.

☻ cherish the power of touch (and pets!)

The Covid-19 pandemic illustrated just how important touch is to many of us. While we could see and hear people over video or on the phone, or wave at a distance in real life, touching those close to us (but who lived in a different household) was out of the question for most. We found we couldn't show affection, console, or comfort people, like we're used to, and during the hardest of times it was unbearable not to be able to hold someone's hand, stroke their brow, or kiss them goodbye. Even individuals allowed to be close to those who were dying were covered head to toe in protective gear. Too many passed away without sensing the reassuring touch of their loved ones or someone who could offer comfort in their final moments. It was the one thing that online creatives and community supporters couldn't fix, despite their eagerness.

Touch hugely affects how we feel. Unsolicited and unwelcome touch can provoke feelings of stress, anxiety, fear, upset, irritation, and more, whereas welcome touch may help us feel calm, welcome, safe, reassured, attractive, and happy, for example. We're all so different, however, that what we view as welcome or unwelcome can only be determined on an individual basis. Some people are very 'touchy-feely' and hug or touch practically everyone, whereas others might feel repelled by such an approach and prefer to maintain a distance and communicate in less tactile ways. And that's fine. It's about understanding how touch can benefit you, within the bounds of your own preferences. So, let's focus on welcome touch.

The sense of touch is one of the first to develop in babies, and touch is hugely important for infant development. In adults, touch is also a core component of social interaction. From a handshake with someone new, a pat on the back from a peer, or a big hug with a good friend or family member, that simple action can instantly stimulate thoughts and feelings within us. Scientists have shown that touch stimulates an increase in oxytocin − a powerful hormone produced in the hypothalamus that is best known for its role in childbirth and breastfeeding. It has other functions, however, including helping us bond with others, fall in love, reduce stress, relax, and alleviate physical pain. Conversely, low levels of oxytocin may be associated with stress and depression.

Oxytocin levels may also be boosted by petting animals. One small study involving Labrador owners found a significant increase in blood oxytocin levels after just a few minutes of sensory interaction with their dogs. Pet ownership has been linked to a range of health benefits, including an increased sense of relaxation and calm. Animal lovers will know that stroking or hugging pets, and interacting with them, is good for wellbeing. The physical act of affectionately touching pets appears to have a similar effect to that seen with other humans, and warm relationships with our pets have been found to lower anxiety, improve mood, prompt mindfulness, and help disrupt rumination and cycles of negative thinking. While the understanding behind how pets benefit our health is still unclear, we know that they can help reduce our blood pressure and lower our heart rate, increase our levels of physical activity, boost happiness, and reduce loneliness through their companionship. At the simplest level, caring relationships with animals can be a welcome bonus to our wellbeing, but for some it can be a vital part of life (particularly for those who don't find it easy to engage with other people or prefer the company of animals). Beyond touch, humans also find watching the antics of animals pleasurable – just think of the thousands of cat videos online or nature documentaries enjoyed by millions.

℮ treat yourself to a massage

A great example of where we can embrace, rather than avoid, sensory input to enhance rest is massage. It's a go-to tool for those seeking relaxation and is synonymous with wellbeing; something that is seen by some as a luxury, and by others as a necessity. But is the massage doing anything useful or are we just enjoying having a lie-down in a soothing environment?

Research into the effects of massage is often complicated by the fact that there are so many different types and innumerable masseurs delivering it, with many individual differences in their application (e.g. hand size and texture, pressure applied, order of techniques, oils used, voice and manner of the practitioner, environmental factors in the location of the massage – temperature, lighting, sounds, furniture, fabrics). However, what research there is points towards typically beneficial effects on health, including pain reduction, relaxation, reduction in stress and anxiety, and improved mood. It seems that massage may stimulate brain activity and reduce our stress response (by increasing parasympathetic activity and lowering cortisol), increase oxytocin levels through touch, as well as reduce inflammation and boost immune function. Less scientifically, lots of people say that it just feels nice. There's no one way to obtain the maximum benefits from massage. You just need to find the right practitioner, delivering your preferred massage style, in the right setting.

A quick note on reflexology is worth a mention here, given its popu-larity in wellbeing literature. Reflexology involves micromovements of the thumbs or fingers on specific pressure points that are said to align with different zones of the body, in particular the organs. While many glowing health claims of the impact of reflexology are made by faithful practition-ers, scientific evidence of its benefits is so far lacking. Some research has found that the experience of having reflexology can be soothing but, on the whole, reviews have shown that much of the research is low quality and no convincing direct effects on health have been found. Any relaxing effects are likely to be the result of the massaging nature of the practice, rather than any specific impact on particular zones of the body.

℮ connect with water

Humans are deeply connected to water (see habit 6 for more on this) and we can choose to interact with it in such a way to enhance our sensory rest. The feel of water on our skin can also spark touch receptors – perhaps due to its temperature, its motion, its direct force, or its effect on our buoyancy – and this may, in turn, lead to positive feelings, like relaxation, con-tentment, exhilaration, and happiness. Imagine luxuriating in a warm bath, or swimming in a clear, calm ocean. The sensation is something we don't just enjoy but often crave. There are also activities that use water to stimulate a natural high (e.g. surfing, deep diving, white-water rafting), and cold-water immersion techniques (such as those practised by extreme athlete Wim Hof) that aim to build mind-over-matter resilience to stress. Ice baths are one thing, but more common is the use of cold showers to draw you into the moment, away from your worries, and to energise.

The thought of a cold shower might make you shudder but bathing in cold water is common in some parts of the world. Whether on its own or combined with heat exposure (such as a sauna), cold-water immersion is claimed to have wide-ranging health benefits, from cardiovascular health and immune-boosting effects to reducing symptoms of mental ill-health. There isn't much scientific evidence to support most of these claims, however. Having tried regular cold showers for a couple of months (during a hot summer – I couldn't manage it in winter!), it is certainly invigorating. It made me feel more alert, fresher, and motivated, and was a great energy and mood boost. Whether it made me any healthier is debatable.

One study, involving over 3,000 adults in the Netherlands, got volunteers to take regular cold showers over three months and compared them to those who didn't take cold showers. The most commonly reported effect was a perceived increase in energy, which participants felt was comparable to the effect of caffeine. They found that the cold shower group had a 30% reduction in sickness absence from work, and the outcome was the same regardless of the length of the cold shower (groups tried showers between 30 and 90 seconds in length). Despite discomfort during the showers, over 90% of participants expressed that they would like to continue cold showers after the trial, as they experienced mood and energy boosts. Whether it is purely psychological or there are physiological actions taking place, cold showers appear to be a useful tool for giving us a lift when we need it. It's also mindful in that it heightens awareness of your senses in the moment. It's a way of breaking out of rumination, as you can't help but think about how cold the water is at the time rather than anything else!

ⓔ reduce environmental sensory stimuli

Another way to exploit the power of water and reduce sensory overload is to try something called flotation therapy. This involves floating in a specially designed shallow pool of highly salted warm water, promoting a sense of weightlessness, in a dark and soundproof room to minimise sensory input. It is said to soothe stress and calm the mind, and has been found in a number of small studies to alleviate symptoms of anxiety and depression as well as alter mood states. It seems that flotation helps people to rest by lowering blood pressure, quietening breathing and heart rate, and encouraging a meditative state where the individual feels more connected with themselves rather than the outside world. Most people who try it seem to enjoy the experience, expressing soothing feelings and happiness, as well as pain relief for arthritis sufferers. While it's something to consider giving a go, it's worth noting that side effects have also been reported, from heightened sensitivity to noise, light, and smell after the session, motion sickness, nausea, and dizziness.

Taking the flotation concept one step further, designer David Wickett has come up with a way of gaining the benefits without the faff of getting wet. David is the inventor of a very special chair, the Elysium, that claims to give us the experience of water-based flotation and send us into a state of deep relaxation. Of course, lots of chairs are intended to be comfortable and restful but this isn't any old comfy chair. It's designed in such a way that it interacts with our slightest movements to send us into a state of weightlessness, akin to the effect of flotation. Stemming from a longstanding interest in the physical relationship and interconnection between the chair and the human body, David has spent years perfecting the technology that enables this unique experience. The result, he says, is that by minimising sensory stimuli on our bodies, we can more easily enter a meditative state.

You first have to learn how to use the chair, as it responds to your slightest movements. Once you're balanced, you experience what we call flotation – a form of true weightlessness where the body moves in the absence of gravity and friction. This, in itself, is restful and removes pressures from the body. But, what's more interesting is that once you're in the flotation, something happens to the brain. It enters a state of both deep relaxation and highly focused attention, much like that achieved through advanced meditation.

David's company (David Hugh) has been working closely with neuroscientists to test the effect of the chair on the brain. Through a combination of EEG and physiological and behavioural studies, they found results similar to those seen in meditation studies. Participants demonstrate increases in measures of mindfulness, and in alpha brain waves, as well as a decrease in cortisol. Measuring how quickly the brain moves into deep relaxation, they found that this occurs in the first five to ten minutes, but after that there is little additional change. This is likely to appeal to anyone interested in experiencing the restful effects of meditation who doesn't have the time or inclination to invest in the lengthy process of learning traditional meditation techniques.

It all sounds amazing, doesn't it?! For most of us, there's a slight snag, however. This is not a product that you'll find widely available. It's a piece of precision engineering combined with high-end design that means it inevitably sits in the luxury market, with a price tag to match (we're talking many thousands of pounds). The good news is that a more affordable option is on the way. At the time of writing, David is working on a new chair aimed at a broader market and he isn't stopping there. A host of related products targeting different aspects of the sensory experience, all with the ultimate aim of deep relaxation with ease, are in the design phase and David hopes they'll find their way to market soon. The purpose is to take a holistic approach to the 'sensory ecosystem', as he refers to it, targeting multiple senses simultaneously to heighten the overall experience. There's still more research to carry out to robustly claim that the Elysium, and other equipment with similar technology, as well as complementary products, can really deliver the same benefits as meditation or flotation, for instance, but the results are promising and causing quite a buzz amongst international investors. Could this approach be the future of furniture and product design? Watch this space!

℮ explore calm technology

Not only are there special chairs available to induce a calm state, but product designers have now recognised the need to quieten technologies. Everyday tech that most of us use is deliberately designed to command our visual and auditory attention, from the beeps and buzzes of smartphone and tablet alerts and notifications, to blinking lights on household appliances as indicators of required action. Just like a small child nagging and nudging for constant attention, technology removes our mental presence from conversations and events by tempting us with latest updates, opportunities to interact, and novel things to watch and do. This information overload is making it difficult for our brains to pick out what's necessary and what isn't, adding to our mental burden and stopping us from being in the moment. It's deliberately manipulating our relationship with the world and the people around us. In effect, this incessant, central focus on technology is stressing us out and removing us from what's really important, rather than aiding us.

Now, obviously, we need technology, and the modern world simply wouldn't function without it, but this doesn't mean that we have to accept the current situation. So, this is where the field of 'calm technology' comes in. There's a new generation of innovators aiming to change things up by placing our interaction with technology at the periphery rather than the centre of our attention. The idea is that technology is to one side, that it can be brought in and out of use as required, rather than always in front of us. Amber Case, author of *Calm Technology: Principles and Patterns for Non-Intrusive Design*, studies the symbiotic interactions between humans and machines and is a high-profile voice on the issue. She explains that, while technology itself is essential, we need to find a balance in how we use it and ensure that we don't allow it to take over our lives. She is keen to emphasise that we should appreciate the value and quality of our time and that technology should generally calmly remain in the background rather than scream for our attention. According to Amber, there are eight core principles of calm technology, including that technology should require the smallest possible amount of attention, it should inform and create calm, and that the right amount of technology is that which is the minimum needed to solve the problem. A basic example of calm technology is an electric kettle. Most of the time we can ignore it as it sits quietly without drawing attention to itself. But it's there when we choose to use it and will alert us when the water is boiling. No more, no less – the minimum needed to do its job. In essence, calm technology should give people what they need to solve their problem and nothing more, to inform without overburdening, and be designed for people first. As Mark Weiser, one of the founders of calm technology theory in the 1990s, once said: "A good tool is an invisible tool. By invisible, we mean that the tool does not intrude on your consciousness; you focus on the task, not the tool."[23] We should regularly remind ourselves that the tech products we use are merely tools, not the focal point of our lives, and we have the power to choose whether, when, and how we use them – it's up to us to prevent them adding to our stress.

23 Weiser, M. *The World is Not a Desktop*. Perspectives article for ACM Interactions. 7 November 1993. https://calmtech.com/papers/the-world-is-not-a-desktop.html (accessed July 2022).

ⓒ limit screen time

It may be some time before we can benefit from calm technology in a big way, but until then we can reduce our interactions with conventional technology to support our sensory rest. With rapid advances in digital technology and access to it, research into its effects on our health has been scaled up. Along with negative effects on our physical health (such as weight gain), there is good evidence linking prolonged screen time with symptoms of depression and poorer quality of life in adults, children, and adolescents. There are also associations with low mood, anxiety, inattention, lower self-esteem, worse sleep, and poorer wellbeing overall.

Research also consistently shows that exposure to screens, such as TVs, smartphones, tablets, and computers, before bedtime significantly impacts sleep. The blue light emanating from such screens suppresses melatonin, which we need to induce and regulate sleep. This light is keeping us artificially stimulated and more awake than we should be. Screen exposure reduces both the quantity and quality of sleep, provoking daytime sleepiness and fatigue, as well as impaired mood, motivation, and attention. So, late night scrolling on your phone or watching TV is something to avoid if you want to get good sleep. Try turning screens off at least one hour before bed and do something more calming instead, like reading a book, doing a calming hobby, or taking a relaxing bath.

The good news is that some technology is proactively reducing the negative impacts from light – e.g. smart lighting that adjusts the wavelength of light according to the time of day, smartphones and tablets that change screen colour during the evening, changes in street lighting designs to direct the light in less harmful ways. There are special lenses for glasses that help minimise blue light from screens and covers you can place in front of computer monitors to do a similar job. But there's still a big issue. The best thing we can do is to unplug – deliberately logging out and disconnecting from technology every day. Not only will this help with light exposure but give us time to be more present in our physical lives.

While getting away from screens is easier said than done, we don't need to let them intrude on every aspect of our day. The internet and social media are major culprits in sucking up our time, but many are now recognising that it's unhelpful for their mental wellbeing. The number of people quitting social media is rising and those who quit report improved life satisfaction. Increasing numbers are also being treated for social media addiction, to enable them to move away from it and change their behaviours to enhance their wellbeing. In short, less time online and more time in the real world can help support sensory rest, as well as other aspects of rest.

◉ slow down

Why are we all rushing through life and not taking time to enjoy the journey? Our inner tranquillity can be improved if we simply slow things down. Yes, we can literally walk and eat more slowly, for instance, but slowing down is about creating more space in our day and reducing unimportant activities that lead to busyness. As we saw in habit 1, to feel truly rested we need to pace our brains appropriately, and slowing down can facilitate this.

We sometimes feel like we're just too busy to think. This reminds me of a very busy friend who recently had a brain fail. When she was at work, she looked at a spreadsheet and her mind suddenly went totally blank. Her thinking mind had ground to a halt. She couldn't comprehend what she was looking at or know what to do. This might be understandable if it was new or unfamiliar, but she had created that spreadsheet herself and is an expert in them. Nothing would come to mind, and it was as if her brain simply would not compute. She was otherwise physically well. This unsettling event led to a week off work and a huge overhaul of her work responsibilities. The main problem was that she had far too much expected of her and is the sort of person who doesn't like to say no, and it finally took its toll – in a very immediate way: "It was such a weird experience to have complete brain exhaustion. It's good for others to be aware that it really does happen!" Her need for a brain break was clear but, while most of us won't experience something quite so dramatic, we all have moments when we seem unable to think clearly. This is a message from our brain to tell us to slow down, even if it's for just a few minutes.

Now, I'm not saying we shouldn't ever be busy, and we all know people who like to be on the go. But this is likely to refer to filling their time with things they enjoy doing or have some control over. It's unlikely to be people who have a million things dumped on them over and above their regular responsibilities, that weigh heavy and take away precious time that they would rather spend with the people or activities they love. Taking time to go slow allows us to observe many sensory happenings we might otherwise miss: the changing colours of the seasons, the chuckles of our children, the smells of dinner cooking on the stove, the feeling of rain or sun on our face, the comfort of fresh bedding. It helps us to focus our energy on satisfying things that nurture the soul, rather than a stream of time thieves – those insignificant activities that eat up our time and steal our joy.

Thinking about this part of the book, I remembered the calming sound of the old-fashioned telephone dial, listening to the soft whir after each number was pulled round. Dial phones were perfect for slowing us down; they couldn't be hurried. The world has since moved on, to push-button phones and touch screens, to make everything quicker and more efficient. But if we think about it, very few calls require us to dial a number fast. How many other things can you think of that we do in a rush but, rationally, don't really need to?

It's when we're juggling several tasks under time pressure that we could probably benefit from slowing down the most. Taking regular micro-breaks to allow your brain to adjust and refocus between each task can make a big difference. For example, when switching between tasks, look out of the window and allow yourself to zone out for fifteen seconds or so; or inhale slowly through the nose for a count of five then slowly exhale through the mouth for another count of five and let your shoulders drop as you breathe out (do this twice); or stop and take a quick stretch. Any brief activity of just a few moments like these allows your brain to metaphorically catch its breath and realign itself ready for the next thing.

what did *the panel* try?

1. Make time for a purposeful period of silence and stillness. This took a little patience for some of the volunteers but led to perceptible calm and relaxation.

I definitely find this sort of thing calming and helpful. Compared to similar things I've done or tried, three minutes definitely felt achievable even on the busiest of days and more likely a habit I could embed and stick to. Not that five to ten minutes is long, but just feels like more of a thing I've got to squeeze in, whereas three minutes felt like a nice little breather, reset, or wind down. Trying to focus also made me realise just how frantic my brain is, but over the week I started to get a bit better at focusing so can see how that would improve over time if I stick with it. I quite often struggle with my mind constantly going ten to the dozen so I'm aiming to try and keep this one up – **Caroline.**

This went perhaps too well. I put out the deck chair in my study with an alarm already set to go and fetch our younger son from school. I looked out of the window, at a fly in the room, heard cars go past, tried to focus on phenomena around me, rather than thoughts and anxieties, fairly successfully I thought, then I closed my eyes and focused on my breathing. I fell lightly asleep. My next attempts also went well. I focused on noises – planes, trains, cars, people on the street. Bit of wandering off thinking about things, but attempting, perhaps with more success, to bring my mind back. I focused a bit more on breathing then closing my eyes. I wonder if sleep should be the outcome of meditation but clearly this routine is very relaxing – perhaps too relaxing! – **Duncan.**

Although taking a period of stillness is a very simple activity, I actually found it challenging. I found that if I didn't do it early in the morning, then I wouldn't manage to get round to it. I need to keep eyes on my youngest child all day long, so once she's up it's hard for me to steal a few relaxing minutes for myself without her. I would say that the more I did this activity, the more I was able to just be in the moment and not make a mental to-do list in my head. I would like to continue trying to do this daily because I think it could be something I enjoy and benefit from – **Meg.**

2. Take a cold shower. Everyone who tried this agreed that, while it was unpleasant during the shower, they felt really energised afterwards. However, none felt they could do this regularly. Perhaps it is one to do now and then when we need to better tune in to our bodies, when we're down or in a rut and need to feel alive again, a quick pick-me-up.

I can honestly say that I never take a cold shower under normal circumstances (usually quite the opposite – I turn the temperature up as high as it goes!) and I was surprised at how different the experience was. I found that I focused on the water streaming down, something that I had never paid attention to before. I would say that it forced me to have some time of stillness as it was impossible to let my mind wander to the thoughts that would usually creep in. In the minutes immediately afterward, I felt invigorated, energised, and more alert. Although once that passed, I did not enjoy the feeling of being chilled. I cannot see this being an activity that I will regularly integrate into my life, but I did find it very interesting to try! – **Meg.**

The cold shower feels so nice afterwards, but I just cannot bear it at the time. I would need someone to force me to stay in for more than ten seconds. I would give it a 1 out of 10 at the time of the shower, but 8 out of 10 for how I felt after! – **Jenny.**

I did this every time I had a shower and was not very keen on it. After getting over the breathlessness (which lasted just a few breaths), I stayed in the cold for maybe 30 seconds. It did make me feel awake and get my circulation going, but I really enjoy a warm shower and get cold easily, so the benefit didn't outweigh the unpleasantness for me – **Julia.**

The idea of getting into a cold shower was just so miserable. It didn't seem like something that would be relaxing at all. But it wasn't so bad once I got in, and I felt incredibly refreshed when I got out! I felt like I had more energy – **Debbie.**

The cold shower was not something I was looking forward to and I only really made a half-hearted attempt. However, as luck would have it, I was having my hair bleached the other day and one of the things they do is wash the bleach out with cold water because it's soothing. I have to report that it really is, and especially if someone else is doing it. So, I would definitely recommend that someone else pouring cold water on your head is a very restful thing to have done – **Anna.**

3. Put something that smells nice on your work desk or somewhere you spend many hours of your day. Participants found that pleasant aromas quickly and easily encouraged sensory rest.

I put a lemon geranium next to me and sometimes rubbed the leaves to get the wonderful lemony smell from it. It was very evocative. I anticipated the smell before I actually released the fragrance. It gave me a moment thinking just about the plant and the smell, even if very briefly – **Julia.**

I was really dubious about this as I don't really like anything smelly really – I never wear perfume or have flowers in the house. I decided to use a lavender candle that I was given as a gift. I was really surprised at how relaxing it was. I kept getting a little waft of the lavender as I worked, which made me look over at the candle and gaze at the flame. When I left the room and came back in, I noticed how the whole room smelt yummy. It just felt nurturing and like I was doing something for myself while I powered through my work. I ordered myself a new candle so that I can have a scent I really like and do this a lot more often. It feels like a little ritual to light it, watch the flame, and wait for the smell. I love being able to look over and see it and smell it every now and then. It's a little moment of being present – **Louise.**

4. Place a restful image or photo, that evokes calm or happy feelings, in the place where you spend many hours of your day. Sometimes it's the simplest of things that can make a big difference to our mood.

I found this the easiest, and the most rewarding, activity. I'd just seen a nice landscape painting in an exhibition so I printed it out and stuck it on my wall. I'm keeping it there. Every time I glance at it, I imagine the sense of being there, and am reminded of holidays abroad to places like Greece. It's great! – **Stuart.**

I chose a photo I took one morning on my run, of the ocean at home as the sun was rising. The colours are beautiful, and it makes me smile as it reminds me of many of my favourite things – water, my home city of Halifax, running etc. I found it calming because it evoked only positive, happy feelings every time I saw it and I would smile inwardly when I caught sight of the image. It was a positive addition to my day to have brief moments of calm unexpectedly worked in. I think this is a brilliant thing to do and I will continue to do it and would absolutely recommend it to others as well – **Meg.**

5. Declutter one small area in your house or workspace. Creating a clear space was felt to be highly beneficial for wellbeing by those who took time to do it.

I decluttered my desk. As I approached it after lunch it was very refreshing to see the clean tabletop. I later made further progress and am quite enjoying the clean feel. I feel more focused when I sit down to work, like there's not a pile of other things I need to do. It also just feels more energising to sit down to a clean desk. Decluttering more areas would certainly help a feeling of being 'on top of things', but I think the hardest part is limiting it to a small area. I find the process of decluttering quite exhausting, actually, although I do feel much better when I've done it. I think I will find another small area to tackle next week – the key is to make it small – **Debbie.**

I decluttered my bedside table and my desk at work on two different days. I found it extremely rewarding and relaxing – **Marta.**

6. Take an online gong bath of at least 20 minutes duration. This activity was tried by participants who hadn't come across it before and they were pleasantly surprised by the restful effects.

As someone who listens to 90s pop music when cooking and anonymous techno beats when running, and otherwise nothing at all, I was wowed by the beauty of the sounds of the gong bath. Who makes this stuff? Even through my crappy headphones it was lovely and transporting. I would love to have felt the vibrations for real. I did it again a few times throughout the week as it was so lovely – **Jenny.**

It took a few tries to find one that I liked. It turns out that gongs are frequently used to set up suspenseful moments in movies, so I kept feeling like something sinister was about to happen! I did finally find one that didn't feel sinister. I found it interesting to pick out the layers of different gong sounds and, surprisingly, I did feel rested afterwards, in a way that just lying quietly in bed during the middle of the day would not feel – **Debbie.**

I already do something similar to help me fall asleep at night. I turn on Classic FM, using the sleep button for about an hour and I'm rarely awake when the radio turns off. It's not quite the same thing, but it works for me – **Paul.**

habit 6:

find space to recharge your ideas for *creative* rest

"We all have the capacity for creative thinking but so-called creative people understand how to free up their minds to allow creativity to happen"
David Wickett (Designer)

Sometimes we get stuck in a rut and our thoughts feel trapped. Maybe we need to come up with a novel idea or find inspiration, but nothing seems to be working. It's like the creative part of the brain has simply stopped and, perhaps, in a way, it has. It's tired and finds it easier to churn out the same ideas on a loop rather than apply effort to come up with something new. It needs a break. Instead of trying to force fresh ideas, we need to give ourselves space to recharge and regenerate, as well as disrupt our usual routines, to release our creative potential. We can attain creative rest by changing things up, removing pressure to think, and freeing our mind to roam as it pleases. *calmism* encourages us to take regular micro-breaks for the mind to avoid overload, but there's much more we can do to help get our thoughts flowing again.

what might help?
⊙ seek a change of scene
As well as making sure the brain gets regular mini-breaks, taking a proper break for your whole self now and then is essential for wellbeing.

Getting away from your home and routines can help disrupt your usual way of thinking and being.

Holidays are the classic way to feel rested and recharged. We look forward to leaving all our worries behind and doing the things we don't get to do the rest of the year, whether that's trying new activities, eating new foods, spending time with loved ones, enjoying the scenery, or simply doing nothing for a change. There's no doubt that much of the time they make us feel better, and research backs this up, showing that getting away from our regular routines and experiencing new things helps the brain acquire much-needed invigoration. We're more relaxed, sleep improves, physical exhaustion is lessened, rumination decreases, stress is reduced, mood is lifted, energy is restored, and social connections are strengthened.

One particularly interesting benefit of holidays is their apparent ability to increase our levels of creativity. Holidays often expose us to new places, sights, sounds, smells, tastes, and experiences, so it's not surprising that this can spark new ideas and inspiration. We can incorporate new insights and think about things from a different angle. Yet, it seems that more may be happening too. Temporarily disengaging from work appears to allow the brain to embed knowledge and generate a period of cognitive recovery that enhances problem-solving upon return to work, presumably much like daydreaming does. This period of mental recovery has also been linked to subsequent increased creativity, but not necessarily originality (i.e. ideas spring to mind more easily but they may be no more new or clever than others). It's as though the brain simply has more energy and flexibility once again to perform at a higher level. So, holidays may help rejuvenate us cognitively, not just physically.

Feeling mentally and physically rested is found to improve job performance and, with evidence of improved creativity too, it shows that taking regular breaks from our routines is important for performing at our best. Some workplaces are better than others at ensuring employees feel able to take time off, and some individuals are better than others at taking time off. There are many cultural (societal or workplace), economic, and practical reasons why people don't take more vacation time, but the bottom line is that working all hours without proper breaks isn't making us better at our jobs, nor is it increasing the effectiveness of our workplace. If you think about it logically, we wouldn't expect an athlete who doesn't take rest, alongside training, to win a world record, so why do we think that relentlessly focusing our brain on work, without giving it time to recuperate, would result in the best output? The science tells us that to have the energy and ideas to work hard and work well, we need to prioritise our health. This will support our stamina, increase motivation and productivity, and enable us to give our best. And this isn't just about the workplace, it applies to any area of our life where we're maxing out. Just as we need to rest our bodies, we also need to rest our minds.

But we can't always be taking holidays and, unfortunately, holidays don't last, and most of us can't get enough of them in the year to keep the brain fresh and creative. While we wouldn't give up our vacation time, we also know that it doesn't have a long-term effect on our rest levels (see habit 1 for more on this). We may feel refreshed for a few weeks afterwards, but it doesn't take long for us to drop back into our regular routines and for our brains to feel exhausted and sluggish again. So, we also need other ways to recharge and rejuvenate our creativity. The best way is to incorporate micro-holidays into our weekly schedules to generate a lasting sense of improved wellbeing. Getting a change of scene can be just the thing to help.

Whether attending an exhibition, visiting a new place, or having a day out to a well-loved location once in a while, going somewhere that breaks you out of your regular routines can revive the mind. The Victorians knew what they were talking about when they coined the proverb 'a change is as good as a rest'. They had observed that having a change could help a person refresh and revitalise. The idea is to get away from your usual environment to shift your thought patterns and sensory experiences, allowing the mind to wander, rest, and recover, and one of the best ways to do this is to get outside.

☻ embrace the outdoors

It has long been known that exposure to natural light is good for us and much research has been carried out into this. One review of 54 different studies, published in 2022 by a team from Tokyo, found that exposure to sunlight is important for wellbeing and improving people's mental health. Time spent in green areas and parks, representing a combination of natural light exposure and physical activity, was of particular benefit. Many of us can also attest to feeling better during the summer months when the days are longer or even when we just spend more time outdoors.

For some, the lack of light in winter months can really bring them down. People with seasonal affective disorder (SAD), a type of depression linked to the seasons, typically experience low mood, irritability, lethargy, sleepiness, and other symptoms during winter months when there is less sunlight and they're more likely to be indoors. The mechanisms behind SAD are thought to involve the lack of sunlight affecting the ability of the hypothalamus to work properly. When this happens, it may interfere with the production of melatonin (hormone that makes you feel sleepy) and serotonin (hormone that affects your mood, sleep, and appetite), as well as disrupt your body's internal clock – a perfect storm that can lead to multiple impacts on a person's wellbeing.

According to US scientists Tracy Bedrosian and Randy Nelson, while we used to shape our days around the availability of sunlight, rising at dawn and retreating to our beds at sundown, the invention of electric light has artificially blurred the boundaries of night and day, affecting our physiological processes in turn. Our bodies were designed to work in synchronisation with the natural daily cycles of light and dark, but this is disturbed by exposure to artificial light at night, leading to health harms. Increased risks of breast cancer and metabolic and psychiatric disorders are some of the harmful effects that have been found, and abnormal circadian rhythms have also been linked to obesity, diabetes, and insomnia. These researchers highlight that most of the human population experiences light pollution, from external streetlights as well as light from electronic devices left on while sleeping (e.g. televisions, computers, digital clocks, even machines in hospital wards). When our bodies no longer get the natural light cues from the environment about when to sleep, our normal circadian rhythms shift out of line, with physical consequences.

As well as exposure to sunlight being essential to health, the warmth of the sun on your back or face also feels good. The warm sensation radiates through the body and makes it smile. We often feel better on a sunny day than a gloomy grey one and our perspective on the world seems a little brighter. But just getting outside, regardless of the weather (unless it's extreme), often improves our mood. A few gulps of fresh air can be refreshing and even contain beneficial compounds for health (see 'surround yourself with trees' below). A large study involving over 35,000 people in the UK found that during the Covid-19 lockdown periods, increased time spent outdoors was associated with a reduction in symptoms of depression and anxiety and an increase in life satisfaction. Much wellbeing research focuses on the impact of being in and around nature, but this study showed that just being outdoors made people feel better. Having said that, the effects were even more marked in those who were more satisfied with their neighbourhood walkability and green spaces.

If you're lucky enough to have a garden, allotment, balcony plants, or even access to a community garden, you may already experience the benefits that spending time in, tending to, and caring for it offers. Gardening has frequently been shown to benefit health. There are many obvious physical benefits to being active, but positive mental health effects also stem from this pastime, including reductions in stress, anxiety, and symptoms of depression, as well as increases in life satisfaction and quality of life. Gardening is frequently listed as people's top preference of hobby for relaxing and it's something that many agree helps them feel a sense of calm and tranquillity. The garden is often seen as a private escape, where you can lose yourself. It's a place to think and simply be. A place where there is no judgement and where the natural world continues to move through the rhythms of life at its own pace. Where the impact of your efforts is visible and where you can enjoy the fruits of your labour (often, quite literally!). Making more time to get into your garden space is a great way to rid yourself of distractions and create mental space.

For the ultimate mindful gardening experience, you could try the practice of Zen gardening. Developed in Japan centuries ago, and popular with samurai warriors to embrace simplicity and self-discipline, it broadly involves a small, minimalist, orderly space featuring carefully arranged rocks, raked sand, water, and pruned bushes and trees. Every part of the garden is thoughtfully placed and styled, and it's meant to be viewed from a particular point. It exudes a sense of beauty, simplicity, and peace, and is often intended for meditation, connecting to nature and to oneself. The sense of rest doesn't come from simply looking at the garden, but from also working in it, from raking the sand into wavy patterns representing water, pruning the vegetation, and being in an orderly space. All these features help to clear mental clutter and free the mind. Most of us are unlikely to have access to such a garden or be able to create one, but you can take some of the principles and apply them to your own spaces. There are even mini desktop Zen gardens you can play with as a way of mindfully zoning out from busy daily life around you. You may prefer to encourage a more organic space to allow wildlife to find its own way and that supports mindful rest as you take time to appreciate the beauty and flow of nature around you.

℮ spend time in nature

Did you know that beyond the four main seasons we're all familiar with, there are 72 micro-seasons in the traditional Japanese calendar that tune in to subtle changes occurring throughout the year? It's a beautiful manifestation of the observation and appreciation of the natural world. These micro-seasons have evocative names, with examples approximately translated as 'fish emerge from the ice', 'first rainbows', and 'dew glistens white on grass'. This teaches us to pay more attention to subtle changes occurring around us. Spending quality time in nature, really noticing what's going on, has been proven, over and over again, to benefit us.

Nature seems to tap into our very core and there's a strong relationship between feeling connected to nature and happiness. One study of over 8,500 people found that those who were more connected to nature were more likely to experience greater positive vitality and life satisfaction. During the pandemic lockdowns, the urge to be outside was never stronger and we craved the natural environment. Parks, woodlands, mountains, and beaches became beacons of sanity, drawing crowds from their homes. The change of scene and connection with nature helped to ground us and restore calm, replenishing our souls at a time of crisis. Some doctors are now even advocating Nature Prescriptions, to connect people with nature in a meaningful way, often tailored to seasonal changes, such as woodland walks, borrowing a dog to walk, making rock patterns on a beach, and birdwatching. Feeling more connected to nature also has benefits for the environment, with those people feeling a greater connection being more motivated to take care of it.

There are a host of potential reasons why nature has health benefits, including the sense of freedom and escape afforded by a quiet, totally natural environment, exposure to clean air, the physical activity it en-courages, social connection with others (we often plan to meet people in natural environments for walks, picnics, play dates), and physiological processes it may stimulate. Being in nature, particularly while walking, can reduce the impact of anxiety and depression, decrease time spent rumin-ating, and lead to improvements in stress levels and sleep, amongst other benefits. This positive impact is greater than either an urban walk alone or experiencing nature without the walking element. Furthermore, the bene-ficial effects have been found to last for several months. Physical activity in green space also appears to yield greater improvements in wellbeing than activity undertaken in either a built environment or indoors.

Hippocrates is credited as once having said: "If you are in a bad mood, go for a walk. If you are still in a bad mood, go for another walk." But how much time should we really be spending in nature to make a meaningful impact on our rest levels? A study using data from almost 20,000 people in England on how they experience the natural environment found that spending at least 120 minutes each week in nature was necessary for good wellbeing, with the peak benefits achieved somewhere between 200 and 300 minutes. This exposure to nature did not need to be taken all at one time but could be spread across the week. That sounds like a lot for busy, working urbanites, for instance, but we could achieve this in a 15-minute daily walk during the week and a longer nature visit on the weekend of an hour and a half to two hours. The research included residential green space as well as other natural environments, so regularly pottering around your local park may well be enough to stimulate benefits. Despite these figures, simply getting outside into nature whenever you can, for however long, can only be good, and, with Mother Nature also being the artist's perfect muse, the potential for creative rest is boundless.

☺ surround yourself with trees

One day whilst writing this book, a radio presenter announced that it was National Tree Day and encouraged us all to stare in wonder at a tree to refresh and restore our energy. It was something that we can all do and, moreover, lots of us are only happy to do. Humans tend to love trees. *The Hidden Life of Trees* by Peter Wohlleben, which compared trees to human families in how they live and care for each other, was an international bestseller as we were keen to discover more about these gentle giants that we so often look to for solace. We love observing, and being around, trees and this also does us good.

A study in Toronto, Canada, found that people who lived in neighbourhoods with a higher density of trees on their streets reported significantly higher health perception and significantly fewer cardio-metabolic problems. The scientists were able to calculate that having ten more trees in a city block improved people's perception of their health by an amount comparable to being seven years younger or having an increase in annual income of $10,000. It's possible that more affluent people, and therefore also those with better health, live in areas of greater tree density, but the study accounted for the median income of the different areas to try to address some social factors that may be in play. Other research indicates that forest environments have an even greater positive impact on our wellbeing than green suburbs.

There are several reasons why forests are so calming and good for wellbeing. First, they present an opportunity to escape the noise, pollution, and sensory overstimulation from urban surroundings. They are often peaceful and allow us to avoid crowds and be on our own if we wish. Forests can also be cool and refreshing, with good air quality, enabling us to breathe more easily. In addition, trees and plants emit biogenic volatile organic compounds (BVOCs), such as phytoncides, that are thought to stimulate physiological changes that benefit our health through anti-inflammatory, antioxidant, and neuroprotective activities. BVOCs, of which there are over 1,000, are released by trees and plants to protect themselves from insects, bacteria, and fungi. Studies have shown that these compounds appear to promote brain function by improving cognitive performance and mood, decreasing mental fatigue, and encouraging relaxation. Forest composition differs so widely across studies that it's difficult to pin down whether it's all wooded areas or specific types of trees and plant species that make the difference to our health. However, a forest environment is certainly a calming one and worthy of further exploration in the pursuit of creative rest.

Shinrin-yoku (or 'wood-air bathing') is a concept that was introduced in the 1980s by the Japanese Ministry of Agriculture, Forestry and Fisheries. The Ministry set up a large research centre into the effects of forest bathing and encouraged the population to give it a go. Its popularity spread to other parts of East Asia, then it gained interest in Scandinavia, with several countries establishing large research projects to investigate the health effects further. In these parts of the world, forest bathing is seen as a form of therapy to promote healing. A key component is to engage our senses while immersing ourselves in nature, and it's often combined with mindfulness, as they share harmonious and symbiotic approaches, to maximise its effects. Along with positive physical effects, such as immunological benefits, reduction in inflammation, and lowered blood pressure, forest bathing has been found to boost mood and reduce stress, anxiety, and symptoms of depression, and improve health-related quality of life.

So, what does forest bathing involve? Taking place over an hour or two, you immerse yourself in nature, soaking in everything it has to offer. You head to a relaxing place full of trees (ideally a forest!) for a leisurely stroll and, as you walk, you're invited to engage your senses, awaken to the here and now, and let your stress float away. When you're next in a wooded area, listen to the sound of leaves rustling in the breeze, raindrops falling, wildlife scampering through the undergrowth, and birdsong floating through the air. Observe the colours, textures, and patterns of nature all around – the trees, plants, paths, water, sky – how the sun glints off different surfaces, the heights and shapes of the trees. Notice the smell of the forest; the damp earth after rain ('petrichor'), the scents of the plants. Physically connect by touching tree bark and leaves, feeling the soft earth or cracking twigs under your feet. Maybe even hug a tree! Breathe slowly and thoughtfully and allow your mind to be in the moment. In that moment, there's no rush, there are no chores, there's nowhere else you must be. Feel the freedom of time and let yourself just be.

Such is the benefit of this approach that forest bathing leader Nicky Dorrington receives referrals from her local health service for people with long Covid, to help manage their symptoms.

Forest bathing sessions run all year round, and observing the changing seasons is a big part of the beauty. Seasons come and go, and this can be a reminder that challenges in life also come and go and change over time. Some weather conditions offer additional sensory experiences and wellbeing benefits. For example, in rainy sessions we have petrichor – the earthy aroma when rain falls on soil – the feel, sound and taste of the rain, and trees also produce more beneficial compounds for us to breathe in.

Forests are perfect environments for reflection. My meditations are centred around the trees that we're with; each have their own unique characteristics yet are connected to the other trees and plants around them as part of a community. Participants are often amazed to learn about the 'wood wide web' – the huge networks of connection under our feet – and once you ponder upon the fact that everything in the natural world is connected, the reflection which often follows is that we, too, are part of nature. This can be an ideal antidote to techno-stress – a rest from the constant demands upon our attention from notifications, emails, social media alerts, and texts. The forest demands nothing from us, and instead offers us a beautiful space to simply slow down and recharge.

Most of us aren't lucky enough to be able to do this daily, or even weekly, but we can take more time to get outside and appreciate nature in a similar way. Find trees near you and spend time close to them; appreciate their form, textures, and scents. What do you notice? We can also bring nature inside to enjoy some of its advantages. You don't always have to go to a forest to inhale beneficial BVOCs as high-quality essential oils also emit some, and placing pot plants and natural materials you collect (e.g. aesthetically pleasing stones, sticks, conkers, leaves, pine cones) provides visual and textural reminders of nature to evoke calm. You can also listen to forest sounds for auditory stimulation, as there are plenty of relaxing albums and apps offering these. As well as getting outside regularly, try to bring the outside in for greater creative rest.

✪ immerse yourself in blue space

As we saw in habit 5, our relationship with water is deep-rooted. Sure, we need to consume it to stay alive, but we are so much more entwined with water throughout our lives. We drink it, bathe in it, swim in it, play with it, watch it, listen to it, sail on it, draw it, visit it, crave it, and more. But why is it so important to how we feel? Why is being in a warm bath so relaxing, or swimming in a clear pool so pleasant, or listening to raindrops gently cascading or a waterfall crashing down so stimulating? Humans' relationship with water is a complex and multifaceted subject. Putting fears around water aside, there's something about water that taps into our sensory needs and desires. We might experience the need to feel refreshed and clean, to feel physically supported and have the weight of gravity on our body lessened, or to have an undisturbed view out to a distant horizon to spark a sense of escape or imagination. Perhaps we desire to feel surrounded by warmth, or feel the soothing, rhythmical motion of the sea, or replace overwhelming urban sights and sounds with beautiful and predictable water scenes and soundscapes. Water soothes us from the stresses and overstimulation of modern life. It also inspires creativity in many. You can't move for drawings, paintings, poems, stories, or even craft projects that have water as their central theme.

Anthropologists believe that connecting with water is hardwired in us. Early humans knew they needed to be near water for survival and to respect the water for their own safety. Clean water sources would have been a primary selling feature of a new habitation site, and not only offered water to drink and cook with, but a geographic tool for locating others (who would also likely settle near the water) and a channel of transportation. It meant that animals were more plentiful as they too would choose to live near water sources. According to researchers, our brains became attuned to the sounds of water as a survival skill, as well as associating positive emotions with it, along with the sight of water.

The beneficial relationship between exposure to blue space (a natural space dominated by water) and wellbeing is well known. Living near the coast, for instance, has been linked with better health, and the presence of freshwater has also been shown to have benefits. Some of this may be due to increased physical activity (e.g. walking) in people who live near blue space, but even accounting for this, health benefits from simply being near blue space remain. In fact, according to a systematic review carried out by the Environment Agency for England and Wales, the majority of visits to blue space are not active. The same review found that people who use blue space, and particularly coastal environments, report feeling restored and happier when they are there. Blue space is associated with appreciating surroundings, longer visits, improvement of mood, and feelings of restoration, to a greater level than green space.

Curiously, exposure to coastal blue space seems to account for most of the positive health effects. While less research into freshwater blue space (e.g. lakes, rivers, canals) has been carried out, the results to date suggest that simply living near freshwater doesn't have the same benefits as living near coastal water, and visiting these places doesn't confer the same level of benefit as visiting the coast. This could be due to many additional sensory aspects of coastal water that may stimulate our brains, including the movement of the water in and out; the waves; the shifting landscape as tides change; the unpredictable nature of the sea, whether gently rippling across the sand or crashing against rocks, for instance; the feel of the ocean as we paddle or swim; the sounds of the sea; and the sensations of the sand beneath our bare feet or through our fingers. It's probable that we engage with the sea in an entirely different way to a river, for example, which we're less likely to go in or touch, hear as many sounds from, or notice big changes in. Also, the sea is the dominant feature by the coast, but with rivers or lakes, it could be green space (such as a woodland, field, or park) dominating the landscape. However, frequently visiting rivers, canals, or the sea has been shown to have a positive impact on mental health.[24]

Blue space can work wonders for replenishing the soul and restoring our energy and appears to win over the benefits of green space, but the

24 It should be noted that most research into the benefits of blue space has taken place in middle to high-income countries, where the coast is often associated with leisure time. In many poorer countries, blue space (such as coastal areas) might be implicated as a source of harm, like flooding, drowning (particularly in countries where many people cannot swim), and disease (e.g. mosquitos around standing water or highly polluted rivers). For populations who face such challenges, blue space may not necessarily be experienced as restful.

coast can be too far away for a lot of people and access to green space is more achievable. We know that spending regular time in green space has its own benefits, so the bottom line is to just try and get into, and engage with, nature as often as possible, wherever that may be.

℮ do something different

As well as physically taking yourself to different places to prompt a mental break and get the creative juices flowing once again, we can transport our minds to somewhere new by disrupting our regular activities.

Trying new things offers stimulation and new insights to the brain. Activities that are novel to you can prompt a sense of heightened awareness and focus, and sharpen parts of the brain, including those involved in learning and memory. Your brain must work harder to think about, and make sense of, what it is being exposed to, giving it a mental workout. Much like going to the gym and applying a variety of exercises to strengthen different muscles, new activities help stimulate the brain to increase its potential. As an example, there is plenty of evidence that people who regularly undertake mentally stimulating activities maintain their cognitive function for longer as they age. Keeping mentally active with new experiences is not only invigorating for our mood and creativity, but it's good for our long-term health too. Trying something new also means you have to concentrate more on what you're doing in the moment; your brain doesn't have time to wander off into other thoughts. So, it anchors you in the present, away from other concerns, and can help encourage internal calm.

Doing new activities and hobbies has been shown to relieve stress and reduce depression. Hobbies are a great way to not only relax but also promote creative rest. They can reduce our mental clutter, enabling greater clarity for creative thoughts, as well as expose us to fresh ideas that may stimulate creative thinking.

When it comes to hobbies, we're deliberately choosing activities we think might interest us and we would enjoy spending our time doing, rather than things we have to do. Because they are hobbies, we don't have to stick with them; we're making our own choices about whether, and how, we engage with them. That element of control is important. So is the aspect of honing in on things that we find pleasurable. Already we're making a decision to take some leisure time doing something positive, which puts us in the right frame of mind to gain benefit. When we do enjoyable activities, this taps into the brain's reward system, creating a feedback loop. We actively choose something we want to do, we enjoy doing it, so our brain tells us to do it more often, consequently increasing the motivation to do it again. This positive feedback loop is a driver for those people who become absorbed in the hobbies that become their passions.

Some activities keep us focused on the moment as we need to concentrate, like reading, arts and crafts, or playing an instrument, and so stop racing thoughts and rumination. Some increase creative rest and wellbeing through interacting with other like-minded people, sparking interesting conversations and ideas, as well as camaraderie. Other activities help us to get in a mental state where we're in a comfort zone or are very alert, like running, swimming, doodling, that frees our mind to daydream or release creative thoughts. There are various hypotheses that when your brain is in a particular state, with certain wavelengths dominating (see habit 1 for more on brain waves), it may facilitate creative thinking. The hard science isn't there yet to prove this one way or the other but it's logical to suppose that being relaxed might give the brain space and time to generate ideas, or when it's in a state of excitement thoughts may come thick and fast.

When seeking creative rest, it can help to try something new for inspiration and to challenge your mind to think from different angles. Disrupting your usual activities might send your ideas in an unexpected direction or help to consolidate, or even discard, those that you weren't initially sure about. Creative writing is a good example. When people experience writer's block or are struggling for inspiration, one suggestion is to put their current ideas to one side and to write about something completely different, typically a topic at random. Flash fiction activities (where you write a brief story on a theme, and of a length, that someone else chooses) aim to spark the imagination by taking some of the freedom away from you – you don't get to decide the topic or length, which can free up your thinking. And because it isn't meant to be your major work, it takes away the pressure to be perfect. It compels your mind to refocus.

I asked creative professional David Wickett what he does to rest and stay creative. He says he feels like he balances his rest well. In fact, so much so, he claims to never yearn for a holiday because he feels like his work is his hobby. He's very much living the oft-quoted phrase: 'choose a job you love, and you'll never have to work a day in your life'. Alongside loving his work, he puts his restful way of life down to a couple of things: physical exercise (particularly running outdoors in nature) and showers! The shower thing seems a bit of a curveball as we all do that, but he finds that he's in such a state of relaxation and autopilot that it frees up his mind to allow creative thoughts to permeate. He has many of his best ideas in the shower, apparently, and wonders whether his mind is entering a similar state to that in the early stages of meditation, which has also been linked with improvements in creative thinking.

David is not alone as many people report having their best ideas in the shower, and this is so common that researchers at the University of Virginia wanted to find out why this might be. We could be tempted to assume that in the shower the brain switches off from its surroundings, subsequently freeing it up to wander and possibly resulting in creative ideas. However, the researchers found that during activities like taking a shower, the brain is still moderately engaged and not meandering without purpose, and it is this that may be stimulating ideas. Perhaps when we're in the shower we're still mentally present enough to shampoo our hair and wash ourselves, and our body is experiencing multiple sensory stimuli, including the temperature and force of the water, the scents of soaps, and the sounds of the water flowing. More is going on than you think. The researchers propose that, in comparison, doing something really boring is less likely to fuel creativity than one in which you are engaged to some extent, even if you're not that focused on it. I guess next time you have a shower you can reflect on what's happening with your thoughts and see if it's a place of inspiration for you. You could, if you're feeling curious, then compare the effect with the experience of having a cold shower, where your awareness is more likely to be focused on immediate physical sensations. How many creative thoughts spring to mind then? My guess is that you might be distracted by other, less creative, attentions!

We don't always feel like trying something new or making time for a hobby when we're busy and stressed. Getting motivated in the first place to do something different can be the main hurdle for some. Mustering the courage to get going might feel like an effort but it's when we're busy that we need it most. The options for hobbies are limitless and choosing something you'll enjoy will keep you more motivated to do it regularly. The key thing about hobbies is that it's not the end point that matters but the process of taking part. Appreciate the journey (however small) and take time to notice the little pleasures along the way. You'll still get to where you're heading but your soul will have got something out of it too.

what did *the panel* try?

1. Spend at least five minutes outdoors observing nature around you.
Participants enjoyed this activity and felt they wanted to do it more often. Sometimes all that's needed is a nudge to go ahead.

I sat on our garden bench one early July evening as it turned to dusk, surrounded by the greenery of our mulberry bush, tomato and courgette plants, the overflowing loganberry bush, with a few flashes of red of the loganberries and the blooming cosmos that I could just spy through the bushes. I could smell the sweet jasmine flowers and I listened to the evening chorus and to the swifts' high-pitched squeaks and watched them circle overhead. Earlier, I stood a metre away from pigeons performing a courting ritual dance with lots of bobbing, turning, and jumping – strangely delicate and elegant considering they are rather lumpy usually! I really enjoyed this and feel I should do it more often – **Graham.**

Having some time outside observing nature was great for me. I haven't been doing this as much without the daily walks since our dog died, so it was nice to take some time to do this on a daily basis. It definitely relaxes me, and I feel better and appreciate the world around me. I think it's a good de-stresser and makes me realise that all my daily worries – mainly work – are so insignificant. I really should keep this one up! – **Liz.**

I looked at the skyline because the view from my desk on the third floor at home provides a wonderful panoramic view of my local environment and, on a practical level, it is very easy to do on a regular basis. I noticed the richness in the colour of the trees, which are incredibly vibrant. These scaled-up florets that are full of leaves are reminiscent of a wild afro and the embodiment of nature coexisting in an urban landscape. Above is the sky that, for the most part, is this uninterrupted vista of blue, punctuated occasionally by an adornment of clouds of different shapes, sizes and patterns. Interestingly, these nondescript patterns begin to morph into identifiable things like animal faces. It's all a little bit crazy but, with cloud gazing, you're never quite sure if your mind is really playing tricks on you. I find looking at the skyline enjoyable probably because it has meditational effects, which help me to take time out of my day so that I can mentally reset. This helps me to be more productive throughout the day. I will certainly continue with this activity partly because it is a by-product of my working from home arrangements, but also because I enjoy this regular dalliance with daydreaming as it helps with my mental wellbeing – **Rob.**

It felt a little bit forced to deliberately go, 'Right, I'm now going to spend five minutes observing nature', especially when I'd like to think that I tune in to nature more casually whenever I'm outside anyway. In contrast, I still really like the calming nature picture that I put up that remains on my wall. It's there whenever I want it; and even when unintentionally looking for a break, and glancing at it even briefly, it still tugs me away towards somewhere pleasant, triggering a little mini escape! – **Stuart.**

2. Get a change of scene and go somewhere different. Breaking away from the usual places helped Panel members rest.

It was my daughter's birthday on Tuesday, and we had a birthday breakfast in a local cafe that my wife and I used to use from time to time. I hadn't been there for ages and, whilst it was busy and noisy, I was able to remember times past as well as times present. I also joined in with a group of 'Friends of Margravine Cemetery', who were accompanying a member of a judging panel for a best kept open space award. Until I became ill, I took part in a lot of their activities, helping to look after the place. It is a very pleasant place to spend time in and it has become a regular strolling area for me and many others, particularly in recent years – **Paul.**

My husband and I were sorting out what needs doing for the rest of the week and then we decided to just head to the beach for a couple of hours. It was great. He says it was the most restful thing he has done in months! – **Anna.**

One afternoon, I went for a walk through central London as part of a charity event and it took me through Hyde Park and around the Serpentine Lake. I was genuinely struck by its beauty; it made me feel like I had been transported to a completely different place, perhaps somewhere in the countryside. Suddenly, I was no longer affected by the intense busyness that is all too common in the London environment. Instead, I felt enveloped by a sense of calmness brought on by the tranquillity of the river and bursts of wildlife in the form of bees humming amongst the flowers or a flock of ducks waddling through the calm lake – **Rob.**

3. Do a hobby of your choice. Hobbies are universally enjoyed and relax the mind as well as create mental space to spark creativity, but we tend to not make enough time for them. Deliberately choosing to do hobbies benefitted Panel members greatly.

I managed some hobbies this week, which was GREAT! Instead of mindless scrolling on my phone when I'm procrastinating, I sat down to read instead – an actual, physical book! I found it to be so much more restful and satisfying than scrolling Instagram or reading the New York Times. *I also read for a shorter time than I might scroll – which is great for not feeling like I'm slacking off during the day. I usually read at night, before bed, but I am going to continue reading during the day for my breaks. I also took a nice long bath this week. I didn't read, just soaked in the warm water and listened to music. It was glorious. I thought I would want to read – I always look forward to reading time – but as soon as I got in, I didn't feel the need to. It was almost like I had too many relaxing things at once and I didn't need it all! I think part of the enjoyment was just having made time to do it and allowing myself that time to enjoy and relax. I'm also hoping to do this more, even though I've never thought of myself as a bath person!* – **Debbie.**

Ruminating is a big issue of mine, so reading is ideal as it engages my mind – **Louise.**

I spent some time out gardening this week – a whole day in the garden on the weekend, plus a couple of hours during the week. Mainly pottering about and tidying, which I love to do! It made me feel happy to have accomplished something as well as generally feel healthier for spending time outside, appreciating the surroundings. Gardening is definitely a good time-out for me, as my mind is focused on this and not any other worries or stresses. I am a fine-weather gardener though, so need to find another suitable hobby for the winter! – **Liz.**

I went swimming at the lido three times, each time for only ten minutes, mostly swimming as fast as I could. It was very invigorating and, though tiring, I felt rested and refreshed afterwards. I also had a couple of runs and it was the same feeling for them – **Graham.**

habit 7:

cherish positive relationships for *social* rest

"Let us be grateful to the people who make us happy
They are the charming gardeners who make our souls blossom"
Marcel Proust

We can be surrounded by people and still feel lonely or have regular social exchanges but still feel a void. Much of our time may be spent with people we don't have a particularly strong connection with but who take up our energy – whether that's work colleagues, school parents, online acquaintances, or casual interactions with delivery people, shopworkers, or neighbours, for instance. The problem is that we're lacking regular meaningful interactions that touch us inside and help us feel connected. To gain social rest, we need to replenish ourselves by spending quality time with the people we really care about and who sustain us. These are the friends and family who listen to us when we need to share, hold us when we need support, lift us when we need a boost, and laugh with us when we want to let our hair down. They allow us to be ourselves and help us find inner calm.

There are also times when the person you really need to spend some time with is yourself, away from others. Taking time away from people allows you to catch your breath, focus on your needs, and reset, and for many of us this can be essential to restoring much-needed energy.

what might help?

☻ prioritise time with good friends and family

If there's one thing we learnt from the pandemic lockdowns, it's how much we missed regular interactions with other people. Sure, at first, for some of us it was liberating to have time to ourselves and a breather in the diary to get on with those things we never had time to do until then. We sorted out all the little DIY jobs that had been waiting, took up hobbies we had been wanting to try, and our baking and language skills came on no end with plenty of opportunity to practise them. But the novelty soon wore off and we found ourselves desperately trying to find ways to keep in touch with others. From video chats and quiz nights to arms-length meet ups at the local park, we craved company. The phrase 'social distancing' came to the fore and it was not seen as a positive move. In short, we needed to be around other people.

Science suggests that humans are indeed hardwired to connect with other humans (even if we don't always feel like it!). In evolutionary terms, it was beneficial to survival to work with others rather than go it alone. Hunting, farming, trading, and caring for others, for example, were all made easier when done in a group, and it was safer to exist together. We also developed advanced vocal, written, and sign language abilities, such was our drive to communicate. While we're no longer hunting and gathering to survive, we retain the need to interact and get along with others and feel socially connected – feeling close and attached in some way – and this is frequently linked with wellbeing.

Denmark is often lauded as one of the happiest countries in the world. There are various cultural reasons for this, but attitudes ingrained in everyday life probably have something to do with it. You have almost certainly come across the Danish concept of *hygge*, which has been talked about at great length in recent years. *Hygge* is all about being convivial with close friends and family in a safe, comfortable, cosy place. It's all about togetherness in a warm atmosphere, and this is something that Danes proudly embrace. There's also another Danish concept that places emphasis here: the concept of *lykke* (mentioned earlier in habit 2). In the six pillars of *lykke* (or happiness), one is 'togetherness'. It seems that spending time with other people makes us happy. The 14th (current) Dalai Lama, head monk of Tibetan Buddhism, would agree. In a televised interview in 2015 he said the best way to fulfil your own happy life is to help others and make more friends, while fellow interviewee South African Archbishop Desmond Tutu believed, in line with the African philosophy of *ubuntu*, that it is because of other people that we open and blossom. *Ubuntu* is the concept that our sense of self is shaped by our relationship with others; 'I am' only because 'we are'.

People have a profound effect on us, even if you think you're not that bothered by them. Have you ever noticed that when you're not actively concentrating on a task, you often think about other people – perhaps a conversation you've had, how you feel about them, or how you feel about yourself in comparison to them? In the workplace, for instance, you may or may not have colleagues who you consider friends, but your relationships with your workmates play a significant part in how you feel about work and how effectively you all work together. When we think about 'toxic' workplaces, this is often due to the culture and people's behaviours rather than just HR policies or working conditions. When you look at this objectively, it seems odd that people who we may not feel close to, or even care about the opinions of, can make us feel so rotten. Conversely, when we experience a kind gesture or friendly response from a stranger, it gives us a little boost inside – why should this be? We don't have any strong feelings about them as we don't know them, and we didn't ask for, or expect, anything from them. So, even if you consider yourself to be someone who is not that interested in connecting with others (or just not a 'people person'), it is still likely that your feelings are influenced by your interactions. And this is all down to being programmed to connect. There's a fascinating book by Matthew D. Lieberman, called *Social*, if you want to read more about this. Lieberman asserts that we're all profoundly shaped by our social environment and that, according to his own research findings, our brains experience physical and social pain in a similar way.

Prisoners are sometimes put into isolation as a punishment, and social isolation has been used as a form of torture because long periods of it can cause significant harm to mental and physical health. Research suggests that people with few, or poor quality, social connections are more likely to suffer from health conditions like heart disease, cancer, and high blood pressure, as well as impaired immune function. This indicates that actively, and positively, interacting with other people is important for our wellbeing. Adults who are socially well connected have been found to live longer than more isolated peers, even after accounting for their socio-economic status and health behaviours. Other research indicates that socialising is also good for brain health. Not only does it lift our mood and help maintain social skills, for example, but there's some evidence that it may improve our memory and cognitive skills, which is why charities working with older people often recommend social activities not only to stave off loneliness but also to maintain thinking abilities.

Because of this innate need to connect, and the importance to our health of doing so, we can enhance our wellbeing by proactively reaching out to say hello. For many of us, there are countless opportunities to connect with people each week – from people we live with or know, to strangers we inevitably interact with. Making an effort to positively engage with people more often can help lift our mood and strengthen our connection with the world at large. This is something positive we can do as we come across people but, while we should embrace these opportunities, we should also channel our energy wisely. Rather than waste energy increasing our engagement in ways that ultimately feel like a burden (e.g. feeling the need to talk to everyone about anything, or accepting every social invitation), social rest may be enhanced by curating and nurturing relationships with those who are most important to us.

We're living at a time when there have never been more opportunities to connect with others and engage with people all over the world, but the irony is that we've also never been so lonely and isolated. Loneliness amongst all ages has been increasing for decades and, although we have ever more sophisticated technologies that enable us to reach people at any time and anywhere, data shows that we're progressively less likely to socialise in real life and more likely to feel a mismatch between desired social relationships and the actual relationships we have. This is leading to life dissatisfaction and reduced mental wellbeing.

Online contact can be helpful for people who are very isolated and lack opportunities, or find it physically or psychologically difficult, to interact with people in person. It also offers a way of communicating and building social connections without the expectations of social etiquette in real life. For example, written messaging allows people to express themselves and find others with mutual interests without worrying about how they look, making appropriate eye contact, or having the most beautiful home. However, data suggests that online social networking is predominantly used to extend and maintain the relationships we already have in real life (such as sharing photos, life updates, stories), and that we may be moving more of these relationships online as we decrease the amount of time we spend participating in real-life interactions with these people. Real-life interactions seem to bring about greater benefits than online interactions, presumably because it is a richer experience, involving more senses and space, for instance. Social cues of touch, movement, smell, body language, and environmental features might all subconsciously affect how we feel about the experience and about the other person or people. To support our social rest, it seems that we should try, where possible, to optimise real-life exchanges with people who are important to us.

Not everyone finds it easy to connect with others but there are techniques that may help. As we've seen elsewhere in this book, various actions have been shown to improve social connection, including better sleep, smiling more, and listening to music. There's also clear science indicating that group music activities, like singing, dancing, and playing an instrument, are great for improving our mood and lowering stress as well as boosting social connectedness. Additional resources elsewhere suggest a host of ways for meeting people and socialising, so look around for ideas if you could do with some help. As having a good social network is strongly linked to wellbeing, making an effort to find ways to positively connect with people is immensely beneficial.

It's also important to recognise the difference between those in our lives who might be considered friendly acquaintances versus those who are genuine friends. You may think you have great online friends, for example, but will they come over when you're ill, or take on your chores when you can't do them, hold your hand when you need it, or still be around for you and wanting to hang out if you decide to use social media less? Or, in real life, perhaps you know a lot of people who you can comfortably do small talk with but couldn't imagine turning to in a time of crisis or sharing your deepest feelings with. Who can you really rely on? Who cares most about you? It's far better to have a small circle of kind and loving friends and family than a large network of 'friends' who you don't have a strong connection with. Choose to channel your energy on the people who really matter rather than amassing lots of peripheral contacts. For social rest and positive mental wellbeing, when it comes to friends it's very much quality, rather than quantity, that matters. Having said that, the quantity of time spent with those close friends and family does matter. Try to devote plenty of intentional, undistracted time (get off your phone!) to the people you care about to maintain strong relationships.

There may be many people in your life who you care about and who you love to be with, but it can sometimes be exhausting. You can find yourself giving too much. To find social rest, you need to spend time with the people who don't leave you feeling drained; the ones who energise and nourish you. Think about people around you who you can really be yourself with and who don't need anything from you. As much as we may love some people, such as a partner, child, parent, or needy friend, they can deplete us of energy by relying on us to provide for them in some way, whether that's through caring responsibilities or emotional needs, for instance. To gain true social rest, we also need people in our lives who offer mutual respect, support, and friendship, without demanding anything particular in return. Close friends who just want to spend time with you are often the path to true relaxation. Find time in your diary regularly to socialise with people like this, even if only for a quick coffee or a chat on the phone. They really will make a big difference to how you feel and take you away from your worries for a while.

spend time away from people

Too much social interaction is not always for the best, of course. So far, I've focused on positive social interactions, but we know that many exchanges can bring us down. Strained social interactions can harm our health, through increasing stress and anxiety as well as inducing poor health behaviours, such as drinking more alcohol, eating less healthily, and smoking, in order to cope. Even minor comments or incidents can leave us feeling tense and irritable and dwelling for days over what's been said or what happened. And simply spending too much time with people, including those we love, can have its downsides too, even when nothing specifically negative has occurred. It can drain our energy levels and we may resent having to spend time in company. Sometimes, we just need a break from people.

Despite everything I just said about the benefits of being with others, there's also a lot to be said for spending some time alone. This isn't the same as cutting yourself off from social connections or becoming a recluse. It's just about taking a bit of time to restore yourself and regain your vitality. After all, when we think about cherishing important relationships, one of those relationships is with ourselves. Taking time for yourself might seem self-indulgent but it is a necessary part of your wellbeing. We can't just keep going, always being busy, always doing things for other people, or in a way they expect, without there being a knock-on effect on our health. Having space to tune in to our own needs, even for just a short time, enables us to carry on with greater energy, clarity, and enthusiasm. Earlier, I talked about mindfulness practices and meditation, and it is when we're alone that we may best try these techniques to calm the mind and body.

How often we need to be in our own company is very much down to the individual. Consider who you know – they will likely be a mix of those who enjoy interacting with big groups and who are drawn to stimulating activities, and those who prefer to spend time with one or two people and enjoy retreating to a quiet space after a while. There is much talk of extroverts and introverts and how they gain their energy, but it appears that those who tend towards introversion can become more easily overstimulated in social situations and have a greater need to withdraw to their own space to recharge. We know that some introverts prefer to avoid social occasions where possible, but many introverts enjoy social-ising. Plenty of introverts can be energised during exciting and interesting social interactions, but they're more likely to feel mentally fatigued after a while and need some time out. At the other end of the scale, even the greatest extroverts need time to themselves now and again. After all, we can't be social butterflies all the time. It can be exhausting to adjust your behaviour to fit in, to fret over whether you're being interesting or clever or funny enough, to stay aware of your body language, to make sure you're talking to the right people at the right time, or to wonder what the right thing to say or do might be. The key here is for each of us to recognise when we need to balance our energy and how some time alone to sit with our thoughts can help us to restore it.

Being alone can be just what you need to replenish and find head space; time to get your thinking in order and settle the mind. It allows you to process your thoughts and feelings, as well as mindfully work with them. As well as facilitating clarity and stimulating creativity, choosing solitude is thought to help us with self-exploration and renewal, and enables us to be more connected with ourselves and feel comfortable in our own presence. This is something that Virginia Woolf certainly appreciated when she wrote:[25] "She could be herself, by herself. And that was what now she often felt the need of – to think; well not even to think. To be silent; to be alone."

Taking positive time away from loved ones may not only benefit us but it may also strengthen our relationships. The pandemic lockdowns showed that we better appreciated the people in our lives after being made to stay away from them and, while we wouldn't want to go through that enforced social distancing again, we can recognise that taking a break may help us better value our relationships. Several studies have even tried to find evidence to back up the famous proverb 'absence makes the heart grow fonder', with some success. Researchers in Finland analysed mobile phone call data from around 400,000 people over a seven-month period and found that when communication becomes less frequent between two people, calls become longer to apparently compensate for this. They speculate that this increased call time is important for reinforcing social bonding and preventing relationships breaking down from lack of interaction. Another study, from the University of Colorado

---

25   In *To the Lighthouse*.

Boulder, examined the brain activity of prairie voles when they were with and apart from their mates. They discovered that when voles had been away from their partner and were then reunited with them, a part of the brain's nucleus accumbens lit up. The nucleus accumbens is associated with rewarding experiences. The longer the animals had been paired, the larger the cluster of cells that was stimulated, suggesting the bond was stronger. Although it's not yet known what that specific cluster of cells does, it looks like mammals are hardwired to invest in developing rewarding relationships with others.

We can, of course, spend too much time alone and in silence (see habit 5 for more about this). This can lead to loneliness, introspection, and too great a focus on our own thoughts, amplifying feelings and falling into negative thinking cycles. But this is about making a choice – choosing solitude when we feel we need it (not having it forced upon us) and using it to rest and replenish. Actively choosing to take time out to simply be yourself, to reflect, as well as to mindfully experience the present, with no external judgement or pressure, is fundamental to achieving social rest.

Even if you have little opportunity to spend time alone, making small adjustments to your lifestyle that better fit with your social interaction preferences and energy needs could make a difference. For instance, creating time between meetings to collect yourself for a few minutes, choosing to attend only some social engagements rather than all, or reducing the length or frequency of regular social meet ups, can all help to reduce the load and recoup some social rest. This might require you to let people know your preferences and needs, which might be uncomfortable at first, but you'll find that learning to say no is one of the most empowering things you can do to help you gain better control of your life and your rest levels.

what did *the panel* try?

1. Spend quality time (either in real life or a phone call) with a friend and tell them how you're feeling. Proactively communicating honestly and openly about feelings with a friend enabled Panel members to gain clarity about themselves as well as develop deeper bonds with people close to them.

I speak to a colleague regularly about work issues and over the past eight months we've developed a friendship that allows us both to talk about things other than work. These conversations take place over video conferencing and usually start off with me complaining about work stuff. This allows me to decompress, so I speak candidly about work-related issues that are causing me frustration. The most gratifying aspect of these conversations isn't the offloading part, although it is satisfying to a degree, but it's listening to his deliberations about work and about his general wellbeing (he has a disability). We talk about family and friends, his medical treatment, and other topics that we share a mutual interest in. Strangely, I feel better as the conversation comes to an end than I do at the outset, perhaps because I am no longer focusing on my own problems, which can be slightly self-absorbing. The gratification comes from being there for someone else and the learning that comes with listening. As a result, my frustrations dissipate, and I feel able to deal with the challenges ahead – **Rob.**

I'm usually pretty open with my friends and often share my feelings, but for this exercise I took it a step further and tried to be totally honest about how I was feeling without watering it down or trying to sugar-coat things. It definitely made me feel better at the time to release some negative feelings that I had building up. However, I would say the downfall was that, because I chose to share feelings of anger and frustration over a situation, my friend felt that she had to make it better for me and try to come up with solutions. I understand this response (and would do the same thing myself in reverse), but I wasn't looking for that, merely just to share my feelings. I will definitely do this more. It feels good to be honest about my feelings and allows me to connect with people on a deeper level – **Meg.**

I didn't get to do this, partly through lack of opportunity, but it did make me think a fair amount about social isolation and my own situation – **Stuart.**

2. Spend intentional time with your children, giving yourself over to be totally present in the moment. For participants with children, this activity felt like being given permission to stop everything else and devote time to simply doing what the kids wanted to do or talk about, without distractions.

Our complete attention is a precious gift that we can all offer, and benefit from, more often.

It sounds terrible, but although I am a stay-at-home mom and therefore am lucky enough to spend an enormous amount of time with my four children, I also have a huge amount of things I need to get done in a day and I hardly ever give the kids my complete and total attention. I'm guilty of often being on my phone texting/sending emails when they are talking to me, or tidying/cooking/ doing laundry, for example, while I'm also trying to juggle playing a game with them, helping with homework, listening to their stories etc. I frequently tell my husband that, despite the fact that I spend all day, every day, with the kids, I am jealous of him. He spends a tiny fraction of the time that I do with them, but I feel his time is of a much higher quality. This activity forced me to spend time with them in a way that I always want to but just don't prioritise because I'm obsessed with being productive and getting things off my to-do list. I loved this activity and felt it lessened my relentless mom-guilt from not giving them enough de- voted attention. I will definitely try to continue doing this on a daily basis, even if it's just for a small amount of time – **Meg.**

3. Dedicate some time to be alone. Few participants felt they had the time to do this, but those who did manage to find solitude for any length of time felt more relaxed afterwards.

I looked forward to this. A solitary evening visit to our allotment to do some watering in lieu of making supper for the boys and being immersed in the near constant bickering and stress of home life was a welcome break. It's a fairly short walk there, but a kind of secret garden once through the gate with few people, by a railway line and not far from a busy road but surrounded by growing plants with low sunlight slanting in from the west. It was certainly a restful, reflective time that was much needed – **Duncan.**

habit 8 :

nurture a sense of purpose and belonging for *spiritual* rest

"I alone cannot change the world,
but I can cast a stone across the water to create many ripples"
Mother Teresa

Up to this point, we've largely focused on what you can do for yourself to feel better. It's been very much of a 'you do you' approach. Now, it's time to consider the importance of thinking about life beyond ourselves. In *calmism*, spiritual rest concerns finding your place in the world and connecting with it more deeply to foster inner calm.

In this age of ambition, it can be all too easy to get wrapped up in ourselves and our own lives. By focusing too much on our immediate priorities and personal goals, we can become a little disconnected from the wider world, living in our own bubble. This might leave us feeling flat and discontented, as though we lack a sense of broader purpose or wonder where we belong. We may feel unquiet and restless, or as though there's an itch we can't quite scratch. A good example is the many wealthy philanthropists who have everything they need but come to realise that something is missing from their lives – they may choose to fill this gap by getting involved with more meaningful, altruistic causes, donating what they can, whether that be money, time, or expertise, for instance. While we may not all be multimillionaires, we can replenish and expand that sense of self through seeking spiritual rest; taking time to consider who we are, as well as taking action to contribute something positive to the world.

what might help?
ⓔ learn from religion and spirituality

There's no doubt that many people find purpose through religion. It's estimated that there are over four thousand recognised religions around the world, each with their own defining features. Amongst this complexity, many major religions share several common elements. Notably, they usually provide a strong sense of community and a feeling of belonging, a sense of a wider purpose in life; they encourage helping others, regularly take time to pause and reflect, and devote dedicated points in the calendar to celebrate. Taking time to rest and reflect is so important that a whole day each week is often devoted to it, like the Sabbath in Jewish and Christian faiths, for instance. These are features that we all need in our life in some way to feel fulfilled, regardless of our individual beliefs.[26] Forgiveness and acceptance are also common aspects that may help individuals move past negative rumination, and different spiritual beliefs tap into the wider sense of self and mindful acceptance.

Architecturally, many religious and spiritual buildings are designed to provide an atmosphere of calm and peace. As well as providing a meeting point for formal worship, they are often a place where people can come and rest quietly with their own thoughts and reflections. Such buildings have traditionally been a place of solace and sanctuary for many people and, for others, they are a centre of joy and celebration. When we think of these places, we might also imagine someone going to pray – a moment to pause, reflect, express gratitude, communicate concerns, and think about something other than oneself, for example.

Praying is so widely practised that scientists have started to investigate what effects it may be having. A 2016 review of twelve randomised

---

26   I should say that none of this negates the fact that some people have had very negative experiences of religion. This discussion is simply to highlight common elements across religions that help people to nourish their whole being.

controlled trials in healthcare settings, involving participants with different faiths, found that praying once or more per day helped people to cope in times of illness and crisis, and to lower anxiety levels. Small brain imaging studies have found similar patterns of activity in those who regularly pray compared to those who regularly meditate. Much like that seen in meditators, an increase in slow, alpha brain waves has been observed in people during prayer, indicating a relaxed state (see 'a note on brain waves' in habit 1) but these findings vary widely across studies and types of prayer involved. It's possible that prayer is acting as a type of meditation, bringing the individual attention into the present moment, although prayer is so diverse a form it's challenging to analyse.

The relatively new scientific field of 'neurotheology' is investigating the link between religion/spirituality and the human brain, to try to explain the neurological basis of any spiritual experiences people may have. Neuroscientist Andrew Newberg, from the University of Pennsylvania, showed that the frontal lobes (which are involved in focused attention) demonstrate increased activity during prayer or meditation. This he considers to be logical as you are concentrating during these occasions. A less obvious finding was that at the same time the parietal lobes markedly decrease their activity. This part of the brain is associated with sensory information (e.g. touch, temperature, pressure) and coordinating that information to help us detect what's happening around us, to align ourselves with that and respond appropriately. Newberg proposes that the decrease in activity in these sensory areas during focused spiritual practices, like meditation, may help to explain the feeling of 'oneness' with the universe and the sensation of losing oneself that some advanced practitioners describe.

Interestingly, religious chanting, which is considered to overlap both meditation and prayer, has been found to reduce mind wandering, increase focus, and promote feelings of tranquillity, and brain changes seen when chanting appear distinct from those found in mindfulness meditation. It seems that the chanting is activating the brain in a different way, indicating that different forms of focused meditative or spiritual practices are producing different patterns of brain activity. Chanting has also been found to stabilise cardiac function and this is thought to contribute to the feeling of emotional calmness.

Taking everything into account, whether you consider yourself to be religious, spiritual, or neither, there are certainly aspects we can draw into our own lives to aid our spiritual rest. Dedicating time each week for quiet reflection, wherever you're able to do so, and considering how you can help others, is a great place to start. Exploring your place in life and what else you can do to connect with the world may also bring greater meaning and life satisfaction.

find purpose in life

Spirituality generally refers to having a sense of, belief in, or connection to something greater than oneself. It's not always something we can explain but may instead be something we feel. Yet we don't always need to turn to spirituality to seek a purpose greater than ourselves. The Japanese concept of ikigai roughly translates as 'reason to live' and is about finding your purpose in life. It's not a religion or spiritual path but a way of understanding what makes your life have worth and what keeps you motivated. Taking time to contemplate your place in the world, what or who are of most value to you, and what you have to offer, is a fundamental part of the process. This doesn't have to be a deep, existential debate leading to a never-ending cycle of questions but can aid us in gaining a little perspective into who we are.

A YouGov survey of over two thousand UK adults in 2021 found that around 60% of people in the general population believed that life has a meaning, and this number rose to 87% among those who are religious. The survey also highlighted that older respondents were more likely to feel that life has meaning than younger respondents. Another survey, carried out by three US universities, investigated the beliefs of people across eight countries[27] and discovered that meaning in life was felt to be a common experience but took effort to attain. Social relationships and happiness were found to be the strongest sources of meaning in life. Interestingly, religious faith did not score highly at all. Of course, measuring what

27 Germany, Japan, Korea, Angola, Norway, Portugal, Singapore, and the United States.

people mean by 'meaning in life' varies widely across surveys and indeed it can mean different things to different people, but I guess it's about what individuals feel is meaningful to them. If I asked you whether you believe that life has a meaning, and what that meaning might be, what would you say? Have you ever taken time to consider this? Perhaps now is that time?!

Realistically, we can't all claim to have been put on this planet to change the world, but we all have meaning in our lives of some sort. It would be great if we were the lynchpin of our community who makes an enormous impact for great numbers of people, but for most of us our value is felt in smaller, less obvious ways. Perhaps we make a valuable contribution at work, are loved and appreciated by our families, help to keep a social group going, or are even adored (or just relied upon) by a beloved pet. For those with few social connections, it may be that they're making a difference when they say hello to a local shopkeeper or delivery person; maybe they make a helpful or warm contribution to an online community or show quiet support to a neighbour. Maybe you're particularly skilled at something or do a valuable task that no one else is prepared to do. There are infinite ways in which your life may have meaning (see also 'reflective recognition' in habit 2). After reflection, you may experience contentment in knowing your value in the world, or you may decide to strive for more. Either way, that's fine. There's no pressure to be anything more than you are or want to be. It's up to you to determine your own purpose or meaning in life. Think about what gets you up in the morning, what you love and what you're good at, as well as what you have to offer the world.

You might feel that achieving success is your purpose but, if you think about it, success is a subjective and individual concept that can change all the time. Chasing success might mean we miss the journey and may never be satisfied – we may never reach our idea of success and feel a constant sense of dissatisfaction or, even if we reach our original target, it may not be what we had hoped for and we then need to strive for more. For complete rest, we should avoid being driven by an unrealistic pursuit of happiness or achievement. Ultimately, it is better to strive for what makes us feel grounded and fulfilled. According to the 14th Dalai Lama, in the televised conversation between himself and Archbishop Desmond

Tutu mentioned earlier, our lives should be meaningful, which will bring us joy, and the ultimate source of a meaningful life is within our own self. Both he and Desmond Tutu agreed that the key to joy is to get in touch with our own natural compassion and to live from there.

ⓔ connect with others and show compassion

If you don't consider yourself religious or spiritual, you might think that you don't really believe in anything beyond yourself or your own experiences, or that you only trust logic and hard facts. However, it's still likely that you share values, beliefs, or principles with other people. Take laws, for instance. While they are merely a social construct, invented by humans, most of us adhere to them, most of the time, and believe that they are important. We know that they are necessary for society to work and trust people around us to behave in the same way. Much has been written about why people follow the law, with suggestions that some do so because they are deterred by the penalties of not doing so, and that some follow the law because they respect legitimate authority. Whatever the reasons, we can also deduce that compliance with laws shows that it is possible to believe in something beyond ourselves that is for the greater good of society. Something as basic and (almost) universal as adherence to laws subconsciously creates a sense of connection with society at large, of shared values and of belonging.

Researchers largely agree that humans have a fundamental need to connect with others and belong, although how this is consistently measured is under debate. A sense of belonging, or being anchored in the world, is a subjective feeling that can be enhanced or hindered by many internal and external factors and perceptions. We may be born into a community or become part of one as we grow. A comment or action by another person may make us feel that we never truly belonged to a community, for example, or perhaps we realise we have different interests or hold converse views after all. On the other hand, unexpected events may draw us into a new community – perhaps a trauma or bereavement leads to a support group that understands how someone is feeling, offering a comforting environment that enables that person to get more involved.

Our sense of place may also change over time, such as shifting friendship groups as our lives change, or moving to a new area. Finding a sense of belonging can take time, but it is there if we look for it.

When Queen Elizabeth II died in September 2022, the UK was thrown into a period of national mourning not seen since the death of Winston Churchill in 1965. Hundreds of thousands of people came onto the streets to pay their respects. This period of mourning revealed a human need to be around others who shared the same sense of loss, those who understood. People could have just watched proceedings on television from home, but they chose to gather outside Royal residences and queue to see the Queen's lying-in-state or watch the funeral procession go by. Others got together in pubs and parks that were showing events on big screens. Watching and experiencing the events together felt important. For those who avidly support a sports team, this behaviour will be familiar; enjoying the camaraderie of showing up to matches with fellow fans, or gathering to watch in sports bars, instead of following the action at a distance on their own. We're also far more likely to laugh if we're in the company of other people than if we're alone as laughter is largely a social emotion. Social context is an important factor in our pleasure. Connecting with others often involves sharing an interest or point of view, but also spending time in their presence.

Sadly, many people struggle to find where they belong, and we're spending less quality time with people we care about and more time alone. In comparison, daily social media use has been increasing year on year, with the average use worldwide in 2022 being at almost two and a half hours per day, according to Statista. Nowadays, many people choose to search for their sense of belonging in the digital world. Online communities have exploded over the years but, while they can sometimes be a genuine source of friendship and support, too often they are filled with casual acquaintances who know nothing about us and fairly hollow interactions – not necessarily the true friends and relationships we would like to find. The thing is, while the internet can be great for reaching people and building a network and 'personal brand', it doesn't always generate anything particularly meaningful or help to forge close and valuable relationships. It is a personal choice whether to engage with it or not, and it can be fun for some. But it can also be a cause of stress and anxiety. As *calmism* emphasises, we can choose to take greater control of our interactions with the world. How we engage with social media is a good example. We can choose to change how we use it, reduce our use, or even stop using it altogether if that would help us reconnect with what matters most to us. In total, we're spending an average of almost seven hours a day online.[28] Is it little wonder we feel we have less and less time to socialise and connect with people in real life? For spiritual rest, it's vital that we don't lose sight of the tangible things that matter, the relationships that are important to us. A few supportive comments online can be a mood boost but can't replace someone offering practical help or a hug when needed or celebrating your successes with genuine delight.

We need to reach out and make time for others to cultivate a sense of belonging. If you're honest, do you really make enough time to spend with the people you care about? Do you find that you often send a quick text rather than pick up the phone, or that immediate priorities and events take over your spare time and many weeks or months pass by before you see family or friends in person? Do you sometimes become so preoccupied with your own thoughts that you're not really listening to someone else's concerns? Do you make time to find out what's important to the people close to you? The bottom line is we must make an effort. Prioritising social rest by cherishing and nurturing significant relationships will also help bolster your spiritual rest as you feel more connected with the people and communities important to you. It's not always going to be us that needs the validation of belonging, our friends and family also need us to be there for them, to show compassion, to physically help or celebrate, to offer a hug, or to simply listen without judgement.

Religions also offer a sense of connectedness and compassion via a common purpose, but we can achieve this elsewhere if we're not inclined towards a particular belief system. Take the #BeKind trend, for example, that gained prominence after the death of UK television presenter Caroline Flack. We were prompted to reflect on our individual actions

---

28   As well as time on social media, this includes internet use, online meetings, gaming etc.

and develop greater compassion for others. It made people stop and think, sparked public debate, and refocused attention on what's important in life. In other words, we were moved to spend more time considering the feelings of other people and our own part in sharing kindness in the world. As a result, we felt connected to a common cause and to others touched by the same message. Nothing about this was religious or spiritual but merely borne out of a facet of human nature – empathy. Whether it's the #BeKind movement or something else, most of us could probably do with expressing greater empathy for others.

Demonstrating compassion can also further reinforce connectedness. Volunteering our time not only helps us express empathy and kindness but is also beneficial to our wellbeing. It makes us happier, extends our social connections, and helps to embed us into our communities (whether they may be geographical or communities of shared interests, for instance). This giving of ourselves connects us to people and issues, and taking time to understand and engage in issues of importance to others can further strengthen our sense of belonging. This can also help build friendships and purpose in life. In *The Art of Happiness: A Handbook for Living*, the 14th Dalai Lama explains that basic spirituality exists outside of religious belief, comprising qualities like goodness, compassion, and caring for others that bring us closer to people. He believes that we can all feel happier and calmer if we exhibit these qualities but, to make this become second nature, it takes time and practice.

℮ perform acts of kindness

As well as showing empathy and expressing compassion towards others, we can go a step further and actively spread kindness. Here, again, the Danish *lykke* pops up. Another critical pillar to this concept of happiness is showing kindness towards others. We can choose to be kind in our approach to life in general but also to carry out random acts of kindness, which have become something of a trend. Random acts could be anything from giving someone a compliment, helping a person carry their luggage up a flight of stairs, volunteering time, telling someone that you appreciate them, or even being extra nice to a supermarket checkout worker who probably gets moaned at a lot! These things often take very little effort but can make a big difference to how the person on the receiving end feels.

While acts of kindness may sound like a mere fashion for the purpose of individual virtue signalling,[29] science suggests that, beyond our egos, it is in fact doing us some good. A study published in 2022, from The Ohio State University, randomly assigned 122 people with elevated anxiety or symptoms of depression to either engage in acts of kindness, social activities, or record their thoughts, over five weeks. All three groups saw improvements to their symptoms and life satisfaction, and these were maintained for at least another five weeks, when measured again. However, the group that saw the greatest benefits in symptom improvement, life satisfaction, and social connection were those carrying out acts of kindness. In addition, a separate small Canadian study assigned three groups of socially anxious people to either acts of kindness, behavioural interventions to reduce negative feelings, or activity monitoring as a control group. Those in the group who performed kind acts showed significant increases in positive feelings, sustained over the four weeks of the study, whereas no increased positivity was found in the other two groups. Participants in the 'kind acts' group also reported an increase in relationship satisfaction and a decrease in social avoidance – an important outcome for individuals with social anxiety. Additional research into kindness towards others has also found positive effects on the happiness and wellbeing of the giver. It seems that this may derive from observing the pleasure or gratitude expressed by the recipient of the kindness, but also from the giver simply knowing that they have done something kind or helpful.

What's particularly amazing about the science behind kindness is not just that it makes us feel better but that it can even result in physical changes in our blood. Sonja Lyubomirsky, behavioural scientist and professor of psychology at University of California Riverside, studies the science of happiness. In one project, her team discovered that participants

---

29 Virtue signalling is conspicuously displaying, or expressing, virtuous views or behaviours to demonstrate one's good character. It is often considered a self-serving and disingenuous action.

who did acts of kindness for others became happier and stayed happier for a number of weeks after the study had finished; people who did acts of kindness for themselves felt good while they were doing it, but it didn't change their happiness levels overall. In other words, the thing that seemed to matter most was how they were connecting with another person. Their investigations went further to then see if this left any physical impression on the body. Blood tests revealed that those who did acts of kindness for others showed changes in gene expression in their blood that are associated with a healthier immune profile. It appears that the happiness and satisfaction that we get from doing something for others is not only benefitting us psychologically but may be giving us a physiological boost too. Happiness has previously been associated with a stronger immune system, as well as increased longevity, and research has also found that just thinking about helping others makes the brain's reward system light up, so perhaps all these things tie together somehow. Whatever is going on in the body, being kind is a positive way to augment our spiritual rest.

Some cultures are naturally more likely to assist people, display kindness, and rally round when someone needs help, although this may be a little more alien in other societies. We often take our cues from people around us in knowing how to behave in any given circumstance, but perhaps this is an area where we shouldn't be afraid to stand out and lead the way. You may live in a busy, hectic city where those in need are ignored and passed by, and where individuals are in too much of a rush and too concerned with their own lives to simply be nice. Just because that's the way others choose to behave, that's not a good reason to follow, to simply keep our heads down and stick to our own business. Every society, no matter how rich or poor, needs kindness and demonstrations of humanity, so why not commit to making your part of the world a little better through kind acts? Perhaps even take a prompt from when someone shows you kindness and pay it forward when you can.

℮ practise and express gratitude

Doing things for other people often highlights the difficulties that some face and leaves us more aware of what we might be thankful for in our own lives. Being grateful for what you have, rather than constantly striving for more, is a key ingredient of spiritual rest. Sometimes we need to accept that we have enough (even if that's just in specific parts of our life) and taking time to appreciate the things that matter most to us can do us some good.

Actively practising gratitude has been consistently shown to improve psychological wellbeing, including improved mood, increased life satisfaction, and reduced anxiety. There's also a link between gratitude and happiness, with those grateful for the simple things in life also appreciative of how others contribute to their happiness. Beyond this, although the evidence is more mixed, there are also glimmers of beneficial effects on physical health too, such as improved energy, better sleep, and lower levels of inflammatory markers. More work is needed to tease out whether these effects really are true, but wouldn't it be fantastic if we could feel better on multiple levels just by reflecting on what we're grateful for?! It certainly seems that gratitude improves wellbeing, and we can all take something from that.

There are many ways to practise gratitude in our routines. We can dedicate time each day or week, for example, to take a moment to think about what we're grateful for and ground ourselves in appreciating what we have. We could choose to simplify our lives by accepting moderation in our needs and wants, for instance. Instead of being swayed by the latest trends, we could ask ourselves whether we have enough clothes and accessories to meet our current needs. Do we really need another gadget or are we fine without it? Is another TV subscription necessary? Feeling envious of other people is also unhelpful to our sense of calm and rest and doesn't get us anywhere. It is generally better for your peace of mind to focus on what you have rather than what you don't have and to give thanks for the important people in your life. As Marcus Aurelius wisely wrote: "Do not indulge in dreams of having what you have not, but reckon up the chief of the blessings you do possess, and then thankfully remember how you would crave for them if they were not yours."

Frequent internal acknowledgement of our gratitude is valuable for our rest levels, but expressing our gratitude and letting people know that we appreciate them is even better. This is a win-win for all. They feel good hearing it and knowing they're important to you, and you feel good also. Making other people happy tends to make us happy too, so why don't we do it more? Societal expectations of behaviour and awkwardness are typically the barriers that stop us, but we can all break out of our comfort zone and tell someone how they've made a positive impact. In the end, we all like to be told that we're appreciated and there's no reason why we can't be the ones to be the bearers of thankful sentiments.

what did *the panel* try?

1. Identify three things you are grateful for, no matter how big or small.
Taking time to reflect on the things that matter and internally express gratitude for them proved beneficial for the spiritual rest of Panel members.

This has been really useful, provided unexpected benefits, and is something I've carried on doing. I've been using an app called '5 Minute Journal' to capture my three things when I wake up in the morning (when I'm still in bed). Some days it has been a challenge to come up with three, but the more you do it, the better you are at recognising things to put in. As I'm very busy with two young kids and running my own business, I often forget what I've written as the day passes. However, what it has done is develop a healthy sense of perspective and a greater sense of calm, a kind of reassurance that things are okay when things happen during the day. It's difficult to articulate, but intentionally writing about the things that are going well or okay means you don't sweat the small stuff so much. It's easy to rush through life and not take a step back to appreciate how things are going – **Garin.**

I firmly believe the science behind this so have been trying to take note of things I am grateful for, for years now. When I try and do it more systematically I end up feeling a bit silly – either it is the same thing every day, the 'big stuff' (home, job, family, animals) or I find myself racking my brains for something new (peony is flowering for the first time, puppy has not peed inside for three days now). Worth doing, worth noticing! But better if you can build a habit of noticing at the time, I suspect. It is definitely calming – **Jenny.**

I am grateful that my daughter can give me so much of her valuable time to make sure her old dad is comfortable. I'm also grateful that everyone I love is fit and well – **Paul.**

I thought about my three things and what they mean to me. Family – my main source of emotional support that is the bedrock to my wellbeing and ensures that there is always a sense of equilibrium in my life from a moral and spiritual perspective. The place where my kids live and go to school – this small coastal town is lovely in the spring/summer/autumn, and it offers them the opportunity to live an outdoor lifestyle that brings them closer to nature and far away from the frenetic and absolutely full-on big city life of London. I visit them regularly, and we often go to the beach to enjoy the expansive horizon, whether it's the North Downs or looking over to the Channel and seeing France in the distance. Lifestyle – I am contented for the most part. I have access to good healthcare, my job and career are progressing relatively well, and my overall level of happiness is good. Obviously, I am also concerned by many things outside of my control, i.e. the things that are a stain on our society like homelessness, the dispossessed, the mental wellbeing of our young people, and caring for the old and most vulnerable members of our society. Their struggles are a constant source of concern and occasionally I find myself feeling very unhappy about society and sad for people directly impacted by these issues. Taking time to be grateful can also be a reminder of how hard others have it – **Rob.**

2. Identify someone who irritates you and think about how they must have their own worries and challenges in life, and how they are human and also need compassion. This may be easier said than done, and Panel members had mixed reactions, but putting yourself in someone else's shoes seems to help reduce rumination about irritations and encourage a calmer mind.

I found the process of empathising with someone that caused me irritation reduced the feelings and reduced the time ruminating about, or focusing on, those feelings and therefore enabled me to move on. I have continued to find this task very helpful in daily life. Taking a step back, breathing, and actively putting myself in the other person's shoes has definitely tempered my reactions and therefore improved my relationships – **Annalise.**

I tried to empathise with someone who irritates me. I'm still irritated with them but at least I found that I stopped thinking about them after the exercise. I think this is a good strategy, but I found it very challenging not to still think: 'I'm right, they're wrong.' As in, yes, I see where they're coming from, but I'm still annoyed with them for being that way! It's probably good that I don't have a lot of people who irritate me right now – **Debbie.**

3. Do something for someone else without any expectation of something in return (i.e. acts of kindness). This activity can feel good when done on your own terms. During my mindfulness course, members were also asked to try this. When asked to do it, it could feel a bit forced but, if you choose yourself to do acts of kindness, it can be uplifting and satisfying.

I do enjoy doing things for other people and I would carry out acts of kindness if the opportunity presented itself, but not actively seek them out or have to think about it. Having it in mind to carry out an act of kindness, small or big, helped me to make space in my mind for it when it was needed and therefore not get stressed by the consequence or impact on my schedule. Doing things for others definitely gives a feeling of connection and a sense of purpose, more than when you are only self-focused – **Annalise.**

embracing *calmism*
in your daily life

"If you want to do it, you can do it. The
question is, do you want to do it?"
Nellie Bly (*Around the World in Seventy-Two Days*)

calmism advocates helping yourself, helping other people, and allowing yourself to be helped, to achieve complete rest. Nailing the eight habits in your daily life will leave you feeling revitalised and closer to the tranquil island within yourself. But how can you put this into practice?

You may think you haven't got time to add anything else into your full schedule. I put it to you that if you have time to spend ten minutes or more in your day to worry or scroll aimlessly through social media or the internet, then you have time to do something more restful instead. Your wellbeing is very important and should be a fundamental part of your routine. Despite being busy, at least twice a day we all brush our teeth without really thinking about it as it's so entrenched in us to keep them healthy. Developing such automatic habits should be the same for our mental and physical wellbeing. Rest habits shouldn't be seen as an added extra but as an essential strategy for health. Adopting *calmism* simply means a commitment to taking control of, and enjoying, your own life, rather than letting things get on top of you. The key is to embed regular rest habits rather than wait until you're completely exhausted before you act.

you're in charge of *you*

I often think that we can learn a lot about how to live a calm and rested life from cats. When they're free to choose, they're great at sleeping, resting, and stretching, are in tune with their senses, take periods of quiet and solitude when they want it, are typically picky about what they eat and drink, and select who they want to spend time with. They are very much experts in their own needs and desires. Deep down, your body and mind also know their needs – it's up to you to really listen to what these are.

You now have the material to be the director of your own life; you're the central character so why not give yourself the best part? You can choose how to respond to the world and how to live in it. The world around you shifts all the time and there will always be things that need doing and people who need you, but you can actively decide how to handle this. If there's just one thing you take from this book, it's to remember that you are in charge of you – you can elect how to be. Be less concerned about what other people think, worry less about trying to meet people's expectations, and forge the path that feels right for you and your wellbeing. I'm not advocating being self-absorbed or inconsiderate as connecting with others, nurturing important relationships, increasing empathy, and compassion are also important elements of rest. It's just that you have one life and you don't know how long it will be, so you must take ownership of your place in the world. If, however, you are really struggling, then it might be time to get professional help.

find what works for *you*

It's not always easy to know what would make you feel better, so take some time to explore the different types of rest and see what might match your needs. We've explored many different philosophies and methods in this book, all of which have something to offer in terms of rest. Devotees of certain approaches might assert that you can't simply pick and choose the best bits but that you need to embrace the whole way of being to benefit. *calmism* takes a different perspective. Here, we ask: why shouldn't you pick and choose the aspects that suit you? After all, it's about finding ways that fit *your* life and *your* goals. While some may choose to throw themselves fully into a strategy (like KonMari, *wabi-sabi*, meditation etc), for many of us this simply isn't practical or desirable and we're not able to dedicate hours of time to it. What's more feasible is to slot in a few quick things that lift us each day or week so that over time we feel a longer-lasting benefit.

I do appreciate, however, that this could all seem a bit disjointed, and it might feel more natural to pursue just one or two aspects of rest more fully, rather than lots of small, unconnected activities. Whatever works for you is fine. You could also try merging activities to gain different types of rest in one go (e.g. healthy sleep habits combined with noting down preoccupations before sleep, or cold showers with mindful thinking, or time in nature while expressing gratitude for what you have, or doing an act of kindness plus a hobby). What matters is that they become a regular part of your life.

approach complete rest a little bit at a time

We all like doing things that result in instant gratification, which is why bad habits are so easy to pick up and so difficult to rid yourself of. Positive habits that reap rewards in the longer term take greater willpower to bed in. Trying to make behavioural changes in our lives is difficult, time-consuming, and/or inconvenient, and typically we either put them off or give up after an initial effort. Here, we can learn from the Kaizen approach to change. While it may sound like an ancient Japanese philosophy, Kaizen was developed in the twentieth century as a management theory that is now widely applied in businesses to strive for efficiency and deliver lasting change. More recently, people have come to realise that some of the core principles of Kaizen can also be used in our daily lives to bring about positive improvements.

Kaizen tells us to make small changes frequently to achieve big transformations. Instead of seeking radical changes to your lifestyle that may not last, making incremental modifications can lead to continuous improvement by gradually building healthy rest habits, while lessening the bad ones. It's about going slowly, making constant progress, no matter how small, and having patience for behaviours to evolve over time. Repeating small actions until you get results and then building from there. It's the repetition that establishes habits. Whilst this requires some dedication, over time the effort required will decrease as behaviours become more entrenched and you hardly think about doing them.

It might be dispiriting if you don't see enormous gains for a while with this approach, and it may be tempting to give up, but remember that you are investing in your wellbeing for the long haul. Some of the activities will take time and practice to deliver noticeable benefits (e.g. meditation, yoga, sleep routines) so don't expect too much too quickly but do persist. Just like when you brush your teeth in the morning you don't notice how healthy they are, but when you're given the all-clear by your dentist several months later you know that the regular brushing and flossing have paid dividends. Every little habit you manage to incorporate into your routine will be contributing in some way to your rest levels – trust that you are moving in the right direction.

Getting started is often the biggest barrier to building new, positive habits so let's make this as easy as possible. According to James Clear, author of *Atomic Habits*, when it comes to developing new habits, something called the 'two-minute rule' is often applied, as any new habit should take less than two minutes to do. You break down your habit goal into easy two-minute chunks. He gives various examples of this including if you want to read more, set yourself a goal of reading one page every night (this seems easy to commit to and you'll likely read more); if you want to run 5k then make it a habit to simply put your running shoes on every evening (and before you know it, you'll be leaving the house and starting to run). Once you've started, it's easier to keep going. He says that any habit can be scaled down into a two-minute version. It sounds so simple, doesn't it?! So, why not ask yourself what you could start doing for two minutes each day that would help you feel more rested in the long term? And once you've got those two minutes under your belt daily, try increasing that to three minutes until that sticks, and then increasingly longer periods (you get the idea). This approach can also be used to diminish bad habits – by trying to reduce them a little bit at a time, and slowly, until it becomes something you hardly notice. It's about gradually breaking brain connections between negative behaviours and feelings, with barely noticeable disruption to your lifestyle.

Ask yourself what's the smallest, easiest action you could take to start to improve your rest? You might think it's trivial and not even worth doing but remember that doing something is better than doing nothing and it's a great way to start and build momentum. A regular stroll around your garden might become a regular stroll around the block and then around a local park and further. Before you know it, you could be walking the length of the Great Wall of China!

make a plan

For each new rest habit you want to embrace, planning how you're going to achieve it will make things easier. Many research studies demonstrate that people who are specific in their plans for new activities are much more likely to do them and stick to them. Planning your approach in advance removes much of the effort of thinking so you can focus on taking action. For instance, if you plan to go for a walk each day, decide what time you will go, where you plan to walk, and for how long. If there's anything you will need (e.g. particular shoes or clothes), or other events you need to move to accommodate the walk, you can get that organised ahead of time too.

Of course, these plans are all well and good but we know what it's like – you plan to do something new three times a week but other things come up that throw those plans out of the window. Maybe you need to work late, you get sick, someone is visiting, your child needs a lift somewhere, and before you know it you've only managed your new activity once or twice in a month and things just slide. It doesn't take much to derail your good intentions as life gets in the way. To avoid disruption, it's a good idea to make a contingency plan so it's easy to implement if things change. Let's say the weather is too bad to get outside or something else needs to be done at your usual walking time, is there another time you could go, or could you replace the walk with an alternative activity? Perhaps you could switch between similar activities as needed to allow for unexpected changes to your routine. For example, you might want to try slow breathing after lunch each day but if you get a cough or cold, that won't be so comfortable, so maybe you could decide to instead have a few minutes of silence sitting with your own thoughts while looking out of the window. The point is to make sure that even when things pop up to disturb your routine, you already have a plan B prepared that you can enact straight away and that will hopefully give you similar rest benefits. It will also help to continue to embed new routines and make it less likely that you'll give up.

Provide yourself with cues to prompt you to do the positive habit. You could set a reminder or alarm, schedule it in your time planner, put a sticky note on a mirror or window, or place any necessary equipment somewhere you can't miss it. You can commit to doing it at the same time each day or week and tell people your intentions to increase your accountability for going ahead. You might also increase your motivation by giving yourself a reward after doing a rest habit (e.g. a nice coffee, an episode of your favourite TV show). Linking new habits with ones you do already is a great way to get them to last. For instance, controlled breathing or tuning into your senses in the moment while you're waiting for the kettle to boil, clearing a surface while waiting for the dinner to cook, a few stretches while watching TV, reflecting on what you're grateful for while vacuuming, or even practising smiling more while in the shower!

appreciate the journey

Every small improvement will get you closer to reaching your long-term goals and appreciating this progress can help to keep you motivated. It can be gratifying to see how far you've come and notice your rest levels re-balancing. You might like to make a quick note of what you've managed to accomplish each week and use this to reflect on your achievements and progress so far, as well as plan your next objective. Set realistic goals and break these down into what you can do and commit to in the next few days, weeks, and months, and the changes will come. Kaizen shows that if you keep making incremental improvements, you don't need to give too much attention to the long-term goal as you should reach it naturally. The British cycling team, under Dave Brailsford, famously achieved enormous success through accumulating marginal gains over time via small incremental improvements. This led to Tour de France wins and Olympic medal hauls. By focusing on each area of rest that we're in need of, and implementing small changes, we may be able to reach our own Olympic heights of energy and inner calm.

Remember: nothing is set in stone. It's your life and your path, so if you try something out and it doesn't feel right, just change direction. Find what works for you and move at your own pace.

choose your rest habits

To aid you in embracing *calmism*, I've summarised the key actions below. The first step is to have a good think about your life and routines and how you feel right now. What aspects of rest might be lacking? You may decide that some rest areas are already well balanced within your current lifestyle, but where else could you make changes to improve your energy levels and inner tranquillity? Keep in mind that we need to satisfy all eight rest types to achieve complete rest.

habit 1: let your brain unwind for *mental* rest

- be mindful
- try meditation
- get organised (enough)
- note your achievements
- take regular breaks (and don't feel guilty about it)
- slow down and look forward to enjoyable events
- acknowledge that you can only do so much

habit 2: process your feelings for *emotional* rest

- recognise and process your emotions
- understand your stress
- cultivate resilience
- accept who you are
- focus on your breathing
- think positively
- smile and laugh more

habit 3: give your body a break for *physical* rest

- understand your sleep
- improve your sleep
- physically rest and relax
- move more

habit 4: nourish from within for *nutritional* rest

- eat enough
- but don't overdo it
- keep your gut happy and it may keep you happy too
- know your body
- minimise stimulating foods and drinks
- try calming foods and drinks
- challenge how you think about food and drink
- try mindful eating

habit 5: tune in to your senses for *sensory* rest

- recognise the impact of noise
- embrace the sound of silence
- try a sound bath
- let the music in
- create a calm personal environment
- cherish the power of touch (and pets!)
- treat yourself to a massage
- connect with water
- reduce environmental sensory stimuli
- explore calm technology
- limit screen time
- slow down

habit 6: find space to recharge your ideas
for *creative* rest

- seek a change of scene
- embrace the outdoors
- spend time in nature
- surround yourself with trees
- immerse yourself in blue space
- do something different

habit 7: cherish positive relationships
for *social* rest

- prioritise time with good friends and family
- spend time away from people

habit 8: nurture a sense of purpose and belonging
for *spiritual* rest

- learn from religion and spirituality
- find purpose in life
- connect with others and show compassion
- perform acts of kindness
- practise and express gratitude

A combination of techniques will help you achieve rest, including reactive tools for use in the moment to respond to an immediate need, as well as pre-emptive behaviours that you practise regularly to manage your rest holistically over time. A good reactive technique for when you're feeling stressed and anxious or your mind is racing is to take a moment to ask yourself, 'What would make me feel one per cent better right now?' and then acting straight away. You identify something small and instantly doable that can help make a little improvement. It could be anything from a hot drink, a couple of minutes by yourself, some fresh air, a hug, or looking at a joyful image. Taking action to feel one per cent better could give you the energy and motivation to take further steps to increase your wellbeing. Each tiny action builds to a bigger impact in the end. It's about halting the downwards spiral of exhaustion and being overwhelmed and climbing your way back up bit by bit.

To proactively develop routine rest habits, look through the following menu of ideas and see what speaks to you. Select a few activities that might help, or that you're at least intrigued enough to try out (check back through the book for more options and details). The frequency of activities is merely a guide. You can do them as often as you like. Suit yourself! Conveniently, many of the activities are relevant to more than one type of rest and, if you can incorporate these into your life, you'll be ticking off several rest areas at once. And they often work synergistically – feeling physically rested and emotionally grounded may help improve your concentration and reduce mental fatigue, for example. Perhaps getting the creative juices flowing once more, by finding space to recharge your ideas, will help your motivation and make you more likely to want to socialise or have the enthusiasm for doing something kind for someone else.

Decide what you can commit to and construct a plan. You don't have to do all the activities at once – some might be daily, others might be weekly or monthly. You might even choose to combine complementary activities as well as switching from bad habits. For instance, instead of looking at your phone as soon as you wake up (which is an immediate source of stress – why do that to yourself?!), take a moment to be mindful: focus on how each bit of you is feeling, take a few controlled breaths, and then, with your morning cuppa, look out of the window and allow your mind to wander. A routine like this may only take a few minutes but will support your mind in starting the day with greater clarity and calm.

You might feel there are too many options and end up being paralysed by indecision and subsequent inaction. When we overthink, we can get stuck. 'Analysis paralysis' is a common issue that most of us experience at some point. My suggestion is that you contemplate which areas of rest are most lacking in your life, then pick one in high need and start with a single, small action focused on that rest area. Give it time to bed in and, when you're ready, add in another action. The most important thing is to simply try something. Activities should feel like a natural extension of your daily life, that build over time and help all the pieces of the rest jigsaw fall into place.

Books like this can often be quite serious and earnest, but one of the best ways to cover multiple rest types is to make time to rejoice in life. Remember to find the fun, appreciate the silly things, and let yourself go once in a while!

menu

MENTAL REST

To-do list   **d or w**
Organise into specific, time-
bound actions and prioritise tasks

Take a break   **d**
Take a 10-minute break
with no distractions

Micro-breaks   **d**
Make space in your diary for
micro-breaks between tasks

Future events   **ar**
Plan things to look forward to

Meditate   **w**
Try a short, guided
mediation to start then
build up to daily practice

EMOTIONAL REST

Question feelings   **ar**
Challenge worries by looking for
facts. What's real or imagined?

Breathe   **d**
Take a few slow, controlled
breaths (at least 5 in and 5 out)

Smile and laugh   **d**
Even if it's fake! Look for positives

Let it go   **ar**
Determine what you can control
and let everything else go

Immediate needs   **d**
Reflect on your feelings and
what you need in the moment

PHYSICAL REST

Sleep   **d**
Establish a healthy sleep
routine and stick to it

Get comfy   **oo**
Create a calm sleep space
to induce relaxation

Daydream   **d**
Allow your mind to wander
for short periods

Exercise   **d**
Whatever activity you
choose, just get moving

Stretch   **d**
Try yoga or 10 minutes
of stretching

NUTRITIONAL REST

Eat regularly   **d**
Don't skip meals, eat enough
but don't overdo it

Limit sugar   **d**
Limit intake of sugary
foods and alcohol

Be gut happy   **d**
Eat gut-friendly foods and avoid
foods that upset your wellbeing

Mindful meal   **m**
Eat a meal mindfully, in silence,
to increase sensory awareness

Drink water   **d**
And lots of it

menu

d = daily   w = weekly   m = monthly   oo = one-off   ar = as required

SENSORY REST

Silence   d or w
Take a period of silence and
stillness to sit with your thoughts

Try a sound bath   oo
Experience the effects of sound
layers resonating through you

Declutter   ar
Create a tidy, calm personal space

Hug   d
Cuddle those close to you
and cuddle your pets

Limit screen time   d
Limit screen use to
what's necessary and do
something restful instead

CREATIVE REST

Change of scene   w or m
Go somewhere different, no
matter how near or far

Get outside   d
Go outdoors and
enjoy the fresh air

Be with nature   d
Observe and appreciate
nature, and get close to trees

Enjoy water   w or m
Spend time in blue space

Hobbies   d or w
Escape your usual routines
with your favourite hobbies

SOCIAL REST

Cherish people   d
Give time and undivided
attention to people you love

Find real friends   ar
Determine your true friendships
and save your energy for them

Be alone   ar
Spend time in your own company

SPIRITUAL REST

Discover you   oo
Consider your place in, and
contribution to, the world

Engage   ar
Reach out to connections and
your community and get involved

Be kind   d
Show empathy and
compassion towards others,
and try acts of kindness

Be grateful   d
Practise and express gratitude

final thoughts from *the panel*

It occurred to me that I have to first identify that I'm NOT in a state of relaxation, in order to feel like I am then IN a state of relaxation or feeling more rejuvenated after doing something to rest. Perhaps it is simply identifying the state that I am in, full stop. I think if I observe that I am not relaxed, then identifying an activity that is relaxing and doing it makes more of a difference than if I just do an activity that generically should be relaxing without thinking about whether I need to relax or not – **Debbie.**

I realised pretty quickly that it's really hard to make consistent time to concentrate on just one thing in my life when I'm in the house. However, one unexpected thing I have carried on with is letting the shower briefly run cold at the end. I just felt like it was something my body wanted to happen – interesting, considering how much I hadn't enjoyed it initially! – **Julia.**

My take-home message from this experiment is that I would like to incorporate nature and hobbies more regularly into my routine but not psychological reflection – **Graham.**

I particularly loved having a candle alight while I was working. It doesn't take any special effort and it doesn't make me feel like something else has been added to my to-do list. It's simple and calming, almost nurturing, and I like that I don't have to put something in my diary, make a special effort, or carve out time. I understand that self-care and rest are not just about simple things like bubble baths, and more about making lifestyle changes, but I think there's a place for the simple things too, that make the ordinary work and the chores that have to be done just a little better – **Louise.**

It certainly makes one think about things. Being outside in nature is always a winner. It's no doubt a good plan to do this more often and tune in to it more. It would be great to have more time for hobbies. Often it seems to be crucial to either get into a routine with something or have something 'to hand' that can easily be picked up whenever there's a spare moment – **Stuart.**

I have a lot of things in my mind that I'm always attempting to juggle, anxiously remembering things that need doing or that might present a problem, or thinking to the future. I very much like the idea of deliberately making space and mindful time in some way, choosing a new route, actually getting out there for a bike ride, simply sitting and reading, and trying to put all else from my mind. I'd say that the morning stretches have been quite an eye-opener and the cold shower is good in the heat, and I've continued with these. I feel they really benefit me. I was also very relaxed when taking five minutes of silence. I would like to bring positive habits back into my life, like running, or even simply walking, writing (for myself), and sober mindfulness. There's a slight air of artificiality about trying to get to grips with new approaches and then practise them in a slightly piecemeal way, but they all seem to lead towards the same essential place of being aware of oneself, making time for oneself and trying to refresh oneself through being present, rather than distraction – **Duncan.**

I'm going to aim for 10–15 minutes where I do just one thing and do it slowly, or have a shorter micro-break where I don't really do anything. I've realised that it's not doing 'nothing', it's just appreciating small moments of the day and really noticing them, rather than letting it all pass by in a hurried blur. I found the activities so beneficial – I hadn't realised how much I try to do several things at once or do things on autopilot (like pick up my phone). It gave me a shock and made me want to change. It was nice to be given permission to do nothing or just concentrate on one thing. Now I've just got to give myself that permission – **Louise.**

It's funny how many of the activities I asked The Panel to try were things they already knew they should do, but what it took to actually do them was simply me asking and them feeling accountable. Often, they felt grateful for being asked to try certain activities, having noticed the benefits, and it was as if they were being given permission. They didn't need my permission, of course, just a gentle nudge. I urge you to give yourself permission to do any activity that will make you feel better. You don't need someone else to tell you to do it, but I'm nudging you right now if that helps!

**It's now over to you. What are you going
to do today, right now, to feel more rested?**

bibliography

calmism: building habits for complete rest

Asp M. Rest: A Health-Related Phenomenon and Concept in Caring Science. *Global Qualitative Nursing Research.* 2015; 2. doi: 10.1177/2333393615583663

Dalton-Smith S. *Sacred Rest: Recover Your Life, Renew Your Energy, Restore Your Sanity.* (Faith Words) 2017

Helvig A., Wade S., Hunter-Eades L. Rest and the Associated Benefits in Restorative Sleep: A Concept Analysis. *Journal of Advanced Nursing.* 2016; 72(1): 62–72. doi: 10.1111/jan.12807. Epub 2015 Sep 15

Ruggeri K., Garcia-Garzon E., Maguire Á., Matz S., Huppert F. A. Well-being is More Than Happiness and Life Satisfaction: A Multidimensional Analysis of 21 Countries. *Health and Quality of Life Outcomes.* 2020; 18: 192. doi: 10.1186/s12955-020-01423-y

Steptoe A., Deaton A., Stone A. A. Subjective Wellbeing, Health, and Ageing. *Lancet.* 2015; 385(9968): 640–648. doi: 10.1016/S0140-6736(13)61489-0. Epub 2014 Nov 6

Williams M., Penman D. *Mindfulness. A Practical Guide to Finding Peace in a Frantic World.* (Piatkus) 2011

habit 1: let your brain unwind

BBC Four. *Mission: Joy – with Archbishop Desmond Tutu and the Dalai Lama.* (viewed May 2022)

Blasche G., Szabo B., Wagner-Menghin M., Ekmekcioglu C., Gollner E. Comparison of Rest-break Interventions during a Mentally Demanding Task. *Stress and Health.* 2018; 34(5): 629–638. doi: 10.1002/smi.2830. Epub 2018 Aug 16

Boccia M., Piccardi L., Guariglia P. The Meditative Mind: A Comprehensive Meta-Analysis of MRI Studies. *BioMed Research International.* 2015; Volume 2015, Article ID 419808, 11 pages. doi: 10.1155/2015/419808

Britton W. B., Lindahl J. R., Cooper D. J., Canby N. K., Palitsky R. Defining and Measuring Meditation-Related Adverse Effects in Mindfulness-Based Programs. *Clinical Psychological Science.* 2021; 9(6): 1185–1204. doi: 10.1177/2167702621996340

Cahn B. R., Polich J. Meditation States and Traits: EEG, ERP, and Neuroimaging Studies. *Psychological Bulletin.* 2006; 132(2): 180–211. doi: 10.1037/0033-2909.132.2.180

Cherry K. *What are alpha brain waves?* Verywell Mind. www.verywellmind.com/what-are-alpha-brain-waves-5113721 (accessed June 2022)

Cramer H., Hall H., Leach M., Frawley J., Zhang Y., Leung B., Adams J., Lauche R. Prevalence, Patterns, and Predictors of Meditation Use Among US adults: A Nationally Representative Survey. *Scientific Reports.* 2016; 6: 36760. doi: 10.1038/srep36760

Farias M., Maraldi E., Wallenkampf K. C., Lucchetti G. Adverse Events in Meditation Practices and Meditation-Based Therapies: A Systematic Review. *Acta Psychiatrica Scandinavica.* 2020; 142(5): 374–393. doi: 10.1111/acps.13225

Gotink R. A., Vernooij M. W., Ikram M. A., Niessen W. J., Krestin G. P., Hofman A., Tiemeier H., Hunink M. Meditation and Yoga Practice are Associated with Smaller Right Amygdala Volume: the Rotterdam Study. *Brain Imaging and Behaviour.* 2018; 12(6): 1631–1639. doi: 10.1007/s11682-018-9826-z

Kane R. *How many people meditate in the world?* Mindfulness Box. 22 April 2022. https://mindfulnessbox.com/how-many-people-meditate-in-the-world/ (accessed June 2022)

Kent L. *The psychology behind to-do lists and how they can make you feel less anxious.* CNN. 14 July 2020. https://edition.cnn.com/2020/07/14/health/to-do-lists-psychology-coronavirus-wellness/index.html (accessed October 2022)

Kral T.R A., Davis K., Korponay C., Hirshberg M. J., Hoel R., Tello L. Y., Goldman R. I., Rosenkranz M. A., Lutz A., Davidson R. J. Absence of Structural Brain Changes from Mindfulness-based Stress Reduction: Two Combined Randomized Controlled Trials. *Science Advances.* 2022; 8(20): eabk3316. doi: 10.1126/sciadv.abk3316

Lambert C. *Shadow Work: The Unpaid, Unseen Jobs that Fill Your Day.* (Counterpoint) 2016

Lardone A., Liparoti M., Sorrentino P., Rucco R., Jacini F., Polverino A., Minino R., Pesoli M., Baselice F., Sorriso A., Ferraioli G., Sorrentino G., Mandolesi L. Mindfulness Meditation is Related to Long-Lasting Changes in Hippocampal Functional Topology during Resting State: A Magnetoencephalography Study. *Neural Plasticity.* 2018; 5340717. doi: 10.1155/2018/5340717

Lee D. J., Kulubya E., Goldin P., Goodarzi A., Girgis F. Review of the Neural Oscillations Underlying Meditation. *Frontiers in Neuroscience.* 2018; 12:178. doi: 10.3389/fnins.2018.00178

Lindahl J. R., Fisher N. E., Cooper D. J., Rosen R. K., Britton W. B. The Varieties of Contemplative Experience: A Mixed-Methods Study of Meditation-Related Challenges in Western Buddhists. *PLOS One.* 2017; 12(5): e0176239. doi: 10.1371/journal.pone.0176239

Love S. *Meditation is a powerful mental tool – and for some people it goes terribly wrong.* 14 November 2018. https://www.vice.com/en/article/vbaedd/meditation-is-a-powerful-mental-tool-and-for-some-it-goes-terribly-wrong (accessed June 2022)

Marriott H. *How to stop to-do lists ruining your life.* Guardian online. 10 August 2015. www.theguardian.com/science/2015/aug/10/how-to-stop-to-do-lists-ruining-your-life (accessed October 2022)

Mayo Clinic. *Meditation: A simple, fast way to reduce stress.* 29 April 2022. www.mayoclinic.org/tests-procedures/meditation/in-depth/meditation/art-20045858 (accessed June 2022)

Microsoft Work Lab. *Research proves your brain needs breaks.* WTI Pulse Report. 20 April 2021. www.microsoft.com/en-us/worklab/work-trend-index/brain-research (accessed July 2022)

Millière R., Carhart-Harris R. L., Roseman L., Trautwein F-M., Berkovich-Ohana A. Psychedelics, Meditation, and Self-Consciousness. *Frontiers in Psychology.* 2018; 9. doi:10.3389/fpsyg.2018.01475

NIH National Center for Complementary and Integrative Health. *Meditation and mindfulness: What you need to know.* June 2022. www.nccih.nih.gov/health/meditation-and-mindfulness-what-you-need-to-know (accessed June 2022)

Omejc N., Rojc B., Battaglini P. P., Marusic U. Review of the Therapeutic Neurofeedback Method Using Electroencephalography: EEG Neurofeedback. *Bosnian Journal of Basic Medical Sciences.*[30] 2019; 19(3): 213–220. doi: 10.17305/bjbms.2018.3785

Pinsker J. *Why your to-do list never ends.* The Atlantic. 15 January 2021. www.theatlantic.com/family/archive/2021/01/to-do-list-tasks-never-end/617674/ (accessed June 2022)

Plebanek D. J., James K. H. *Why handwriting is good for your brain.* Frontiers for Young Minds. https://kids.frontiersin.org/articles/10.3389/frym.2022.623953 (accessed October 2022)

Powell A. *When science meets mindfulness.* The Harvard Gazette. 9 April 2018. https://news.harvard.edu/gazette/story/2018/04/harvard-researchers-study-how-mindfulness-may-change-the-brain-in-depressed-patients/ (accessed June 2022)

Robson D. *Interoception: the hidden sense that shapes wellbeing.* Guardian online. 15 August 2021. www.theguardian.com/science/2021/aug/15/the-hidden-sense-shaping-your-wellbeing-interoception (accessed June 2022)

Scullin M. K., Krueger M. L., Ballard H. K., Pruett N., Bliwise D. L. The Effects of Bedtime Writing on Difficulty Falling Asleep: A Polysomnographic Study Comparing To-Do Lists and Completed Activity Lists. *Journal of Experimental Psychology: General.* 2018; 147(1): 139–146. doi: 10.1037/xge0000374

Simon S. *Too much mindfulness can worsen your mental health.* Verywell Health. 2 June 2021. www.verywellhealth.com/mindfulness-can-be-harmful-researchers-say-5186740 (accessed June 2022)

Sohal M., Singh P., Dhillon B. S., Gill H. S. Efficacy of Journaling in the Management of Mental Illness: A Systematic Review and Meta-analysis. *Family Medicine and Community Health.* 2022; 10(1): e001154. doi: 10.1136/fmch-2021-001154

Tang Y-Y., Hölzel B., Posner M. The Neuroscience of Mindfulness Meditation. *Nature Reviews Neuroscience.* 2015; 16: 213–225. doi: 10.1038/nrn3916

Thorpe M. *12 Science-based benefits of meditation.* 27 October 2020. www.healthline.com/nutrition/12-benefits-of-meditation (accessed June 2022)

---

30   Now known as *Biomolecules & Biomedicine,* as of January 2023.

habit 2: process your feelings

Baikie K. A., Wilhelm K. Emotional and Physical Health Benefits of Expressive Writing. *Advances in Psychiatric Treatment.* 2005; 11(5): 338–346. doi: 10.1192/apt.11.5.338

Brackett M. *Permission to Feel.* (Celadon Books) 2019

Bravata D. M., Watts S. A., Keefer A. L., Madhusudhan D. K., Taylor K. T., Clark D. M., Nelson R. S., Cokley K. O., Hagg H. K. Prevalence, Predictors, and Treatment of Impostor Syndrome: a Systematic Review. *Journal of General Internal Medicine.* 2020; 35(4): 1252–1275. doi: 10.1007/s11606-019-05364-1. Epub 2019 Dec 17

Carmeli A., Yitzhak-Halevy M., Weisberg J. The Relationship Between Emotional Intelligence and Psychological Wellbeing. *Journal of Managerial Psychology.* 2009; 24(1): 66–78. doi: 10.1108/02683940910922546

Chapman B. P., Fiscella K., Kawachi I., Duberstein P., Muennig P. Emotion Suppression and Mortality Risk Over a 12-year Follow-up. *Journal of Psychosomatic Research.* 2013; 75(4): 381–385. doi: 10.1016/j.jpsychores.2013.07.014

Coles N. A., March D. S., Marmolejo-Ramos F. et al. A Multi-lab Test of the Facial Feedback Hypothesis by the Many Smiles Collaboration. *Nature Human Behaviour.* 2022; 6: 1731–1742. doi: 10.1038/s41562-022-01458-9

Cousins L. *Can always staying positive be bad for our health?* HCF. August 2022. www.hcf.com.au/health-agenda/body-mind/mental-health/downsides-to-always-being-positive (accessed September 2022)

Cross M. P., Acevedo A. M., Leger K. A., Pressman S. D. How and Why Could Smiling Influence Physical Health? A Conceptual Review. *Health Psychology Review.* 2022. doi: 10.1080/17437199.2022.2052740

Dejonckheere E., Bastian B., Fried E. I., Murphy S. C., Kuppens P. Perceiving Social Pressure Not to Feel Negative Predicts Depressive Symptoms in Daily Life. *Depression and Anxiety.* 2017; 34(9): 836–844. doi: 10.1002/da.22653. Epub 2017 May 12

Ferstl M., Teckentrup V., Lin W. I., Kräutlein F., Kühnel A., Klaus J., Walter M., Kroemer N. B. Non-invasive Vagus Nerve Stimulation Boosts Mood Recovery After Effort Exertion. *Psychological Medicine.* 2022; 52(14): 3029–3039. doi: 10.1017/S0033291720005073. Epub 2021 Feb 15

Field B. *The power of future thinking for healthy living.* Verywell Mind. 8 June 2021. www.verywellmind.com/the-power-of-future-thinking-5114362 (accessed October 2022)

Ford B. Q., Shallcross A. J., Mauss I. B., Floerke V. A., Gruber J. Desperately Seeking Happiness: Valuing Happiness is Associated with Symptoms and Diagnosis of Depression. *Journal of Social and Clinical Psychology.* 2014; 33(10): 890–905. doi: 10.1521/jscp.2014.33.10.890

Jaret P. *The surprising benefits of stress.* Greater Good Magazine. 20 October 2015. greatergood.berkeley.edu/article/item/the_surprising_benefits_of_stress (accessed September 2022)

Jibeen T. Unconditional Self Acceptance and Self Esteem in Relation to Frustration Intolerance Beliefs and Psychological Distress. *Journal of*

Rational-Emotive and Cognitive-Behavior Therapy. 2017; 35: 207–221. doi: 10.1007/s10942-016-0251-1

Kim E. S., Hagan K. A., Grodstein F., DeMeo D. L., De Vivo I., Kubzansky L. D. Optimism and Cause-Specific Mortality: A Prospective Cohort Study. *American Journal of Epidemiology.* 2017; 185(1): 21–29. doi: 10.1093/aje/kww182

Littlefield C. *A better way to recognize your employees.* Harvard Business Review. 25 October 2022. hbr.org/2022/10/a-better-way-to-recognize-your-employees (accessed October 2022)

Louie D., Brook K., Frates E. The Laughter Prescription: A Tool for Lifestyle Medicine. *American Journal of Lifestyle Medicine.* 2016; 10(4): 262–267. doi: 10.1177/1559827614550279

Ma X., Yue Z-Q., Gong Z-Q., Zhang H., Duan N-Y., Shi Y-T., Wei G-X., Li Y-F. The Effect of Diaphragmatic Breathing on Attention, Negative Affect and Stress in Healthy Adults. *Frontiers in Psychology.* 2017; 8: 874. doi: 10.3389/fpsyg.2017.00874

Magnon V., Dutheil F., Vallet G. T. Benefits from One Session of Deep and Slow Breathing on Vagal Tone and Anxiety in Young and Older Adults. *Nature Scientific Reports.* 2021; 11: 19267. doi: 10.1038/s41598-021-98736-9

Mayo Clinic. *Stress relief from laughter? It's no joke.* 29 July 2021. www.mayoclinic.org/healthy-lifestyle/stress-management/in-depth/stress-relief/art-20044456 (accessed August 2022)

Mora-Ripoll R. The Therapeutic Value of Laughter in Medicine. *Alternative Therapies in Health and Medicine.* 2010; 16(6): 56–64. PMID: 21280463

Mund M., Mitte K. The Costs of Repression: A Meta-analysis on the Relation Between Repressive Coping and Somatic Diseases. *Health Psychology.* 2012. 31(5): 640–649. doi: 10.1037/a0026257. PMID: 22081940

Ranzijn R., Luszcz M. Acceptance: A Key to Wellbeing in Older Adults? *Australian Psychologist.* 1999; 34: 94–98. doi: 10.1080/00050069908257435

Romundstad S., Svebak S., Holen A., Holmen J. A 15-Year Follow-Up Study of Sense of Humor and Causes of Mortality: The Nord-Trøndelag Health Study. *Psychosomatic Medicine.* 2016; 78(3): 345–353. doi: 10.1097/PSY.0000000000000275

Tamir M., Ford B. Q. Should People Pursue Feelings that Feel Good or Feelings that Do Good? Emotional Preferences and Well-being. *Emotion.* 2012; 12(5): 1061-1070. doi: 10.1037/a0027223. Epub 2012 Feb 6

Trick L., Watkins E., Windeatt S., Dickens C. The Association of Perseverative Negative Thinking with Depression, Anxiety and Emotional Distress in People with Long Term Conditions: A Systematic Review. *Journal of Psychosomatic Research.* 2016; 91: 89–101. PMID: 27894469

Tugade M. M., Fredrickson B. L. Resilient Individuals Use Positive Emotions to Bounce Back from Negative Emotional Experiences. *Journal of Personality and Social Psychology.* 2004; 86(2): 320–333. doi: 10.1037/0022-3514.86.2.320

Marchant N. *Analysis: Negative thinking linked with more rapid cognitive decline, study indicates.* UCL News. 12 June 2020. www.ucl.ac.uk/news/2020/jun/analysis-negative-thinking-linked-more-rapid-cognitive-decline-study-indicates (accessed October 2022)

Wang L., Wang Y., Wang Y., Wang F., Zhang J., Li S., Wu M., Li L., Rong P. Transcutaneous Auricular Vagus Nerve Stimulators: A Review of Past, Present, and Future Devices. *Expert Reviews of Medical Devices*. 2022; 19(1): 43–61. doi: 10.1080/17434440.2022.2020095. Epub 2022 Jan 13

Wiking M. *The Little Book of Hygge*. (Penguin Life) 2016

Wim Hof. www.wimhofmethod.com (accessed October 2022)

habit 3: give your body a break
Basso J. C., Suzuki W. A. The Effects of Acute Exercise on Mood, Cognition, Neurophysiology, and Neurochemical Pathways: A Review. *Brain Plasticity*. 2017; 2(2): 127–152. doi: 10.3233/BPL-160040

Dutheil F., Danini B., Bagheri R., Fantini M. L., Pereira B., Moustafa F., Trousselard M., Navel V. Effects of a Short Daytime Nap on the Cognitive Performance: A Systematic Review and Meta-Analysis. *International Journal of Environmental Research and Public Health*. 2021; 18(19): 10212. doi: 10.3390/ijerph181910212

Gothe N. P., Khan I., Hayes J., Erlenbach E., Damoiseaux J. S. Yoga Effects on Brain Health: A Systematic Review of the Current Literature. *Brain Plasticity*. 2019. 5(1): 105–122. doi: 10.3233/BPL-190084

Hackford J., Mackey A., Broadbent E. The Effects of Walking Posture on Affective and Physiological States During Stress. *Journal of Behavior Therapy and Experimental Psychiatry*. 2019; 62: 80–87. doi: 10.1016/j.jbtep.2018.09.004. Epub 2018 Sep 17

Healthy Sleep. Harvard Medical School. *Sleep, Performance and Public Safety*. 18 December 2007. healthysleep.med.harvard.edu/healthy/matters/consequences/sleep-performance-and-public-safety (accessed October 2022)

Kings College London. *Sleep deprivation may cause people to eat more calories*. 2 November 2016. www.kcl.ac.uk/archive/news/kings/newsrecords/2016/11%20november-/sleep-deprivation-may-cause-people-to-eat-more-calories (accessed October 2022)

Lamp A., Cook M., Soriano Smith R.N, Belenky G. Exercise, Nutrition, Sleep, and Waking Rest? *Sleep*. 2019; 42(10): zsz138. doi: 10.1093/sleep/zsz138

Landolt H-P., Holst S. C., Sousek A. (2014). Effects of Acute and Chronic Sleep Deprivation. In: Bassetti, C., Dogas Z., Peigneux P. *Sleep Medicine Textbook*. Regensburg: European Sleep Research Society. 2014; 49–61. www.zora.uzh.ch/id/eprint/107182/

Linden D. *The truth behind the 'runners high' and other mental benefits of running*. Johns Hopkins Medicine. www.hopkinsmedicine.org/health/wellness-and-prevention/the-truth-behind-runners-high-and-other-mental-benefits-of-running (accessed October 2022)

Nair S., Sagar M., Sollers J. 3rd, Consedine N., Broadbent E. Do Slumped and Upright Postures Affect Stress Responses? A Randomized Trial. *Health Psychology*. 2015; 34(6): 632–641. doi: 10.1037/hea0000146. Epub 2014 Sep 15

National Institute for Safety and Occupational Health. *NIOSH training for nurses on shift work and long work hours*. Centers for Disease Control and Prevention. 31 March

2020. www.cdc.gov/niosh/work-hour-training-for-nurses/longhours/mod3/08.html (accessed October 2022)

Reddy S., Reddy V., Sharma S. Physiology, Circadian Rhythm. [Updated 2022 May 8]. In: *StatPearls [Internet]*. Treasure Island (FL): StatPearls Publishing. Jan 2022. www.ncbi.nlm.nih.gov/books/NBK519507/

Sharma A., Madaan V., Petty F. D. Exercise for Mental Health. *Primary Care Companion to the Journal of Clinical Psychiatry*. 2006; 8(2): 106. doi: 10.4088/pcc. v08n0208a

University of Surrey. *Lack of sleep alters human gene activity*. 11 March 2013. www. surrey.ac.uk/features/lack-sleep-alters-human-gene-activity (accessed October 2022)

Wamsley E. J. Memory Consolidation During Waking Rest. *Trends in Cognitive Science*. 2019; 23(3): 171–173. doi: 10.1016/j.tics.2018.12.007. Epub 2019 Jan 22

Waters F., Chiu V., Atkinson A. and Blom J. D. Severe Sleep Deprivation Causes Hallucinations and a Gradual Progression Toward Psychosis with Increasing Time Awake. *Frontiers in Psychiatry*. 2018; 9: 303. doi: 10.3389/fpsyt.2018.00303

Wilkes C., Kydd R., Sagar M., Broadbent E. Upright Posture Improves Affect and Fatigue in People with Depressive Symptoms. *Journal of Behavior Therapy and Experimental Psychiatry*. 2017; 54: 143–149. doi: 10.1016/j.jbtep.2016.07.015. Epub 2016 Jul 30

Worley S. L. The Extraordinary Importance of Sleep: The Detrimental Effects of Inadequate Sleep on Health and Public Safety Drive an Explosion of Sleep Research. *Pharmacy and Therapeutics*. 2018; 43(12): 758–763. PMID: 30559589

habit 4: nourish from within

British Dietetic Association. *Food and mood: Food Fact Sheet*. August 2020. www.bda. uk.com/resource/food-facts-food-and-mood.html (accessed September 2022)

Huang Q., Liu H., Suzuki K., Ma S., Liu C. Linking What We Eat to Our Mood: A Review of Diet, Dietary Antioxidants, and Depression. *Antioxidants*. 2019; 8(9): 376. doi: 10.3390/antiox8090376

Hulsken S., Märtin A., Mohajeri M. H., Homberg J. R., Food-Derived Serotonergic Modulators: Effects on Mood and Cognition. *Nutrition Research Reviews*. 2013; 26(2): 223–234. doi:10.1017/S0954422413000164

Jenkins T. A., Nguyen J. C. D., Polglaze K. E., Bertrand P. P. Influence of Tryptophan and Serotonin on Mood and Cognition with a Possible Role of the Gut-Brain Axis. *Nutrients*. 2016; 8(1): 56. doi: 10.3390/nu8010056

Lallanilla M. *How junk food makes a bad mood worse*. Live Science. 18 March 2013. www.livescience.com/27977-junk-food-bad-mood.html (accessed September 2022)

Lee H-S., Chao H-H., Huang W-T., Chen S. C-C., Yang H-Y. Psychiatric Disorders Risk in Patients with Iron Deficiency Anemia and Association with Iron Supplementation Medications: A Nationwide Database Analysis. *BMC Psychiatry*. 2020; 20: 216. doi: 10.1186/s12888-020-02621-0

Mantantzis K., Schlaghecken F., Sünram-Lea S. I., Maylor E. A. Sugar Rush or Sugar Crash? A Meta-Analysis of Carbohydrate Effects on Mood. *Neuroscience & Biobehavioural Reviews*. 2019; 101: 45–67. doi: 10.1016/j.neubiorev.2019.03.016

Naidoo U. *Nutritional strategies to ease anxiety.* Harvard Health Publishing. 28 August 2019. www.health.harvard.edu/blog/nutritional-strategies-to-ease-anxiety-201604139441 (accessed October 2022)

Naidoo U. Eat to Beat Stress. *American Journal of Lifestyle Medicine.* 2020; 15(1): 39–42. doi: 10.1177/1559827620973936

NHS. *What is the glycaemic index?* 17 June 2022. www.nhs.uk/common-health-questions/food-and-diet/what-is-the-glycaemic-index-gi/ (accessed September 2022)

Ramirez J., Guarner F., Bustos Fernandez L., Maruy A., Sdepanian V. L., Cohen H. Antibiotics as Major Disruptors of Gut Microbiota. *Frontiers in Cellular and Infection Microbiology.* 2020; 10: 572912. doi: 10.3389/fcimb.2020.572912

Reyes T. M. High-Fat Diet Alters the Dopamine and Opioid Systems: Effects Across Development. *International Journal of Obesity Supplements.* 2012; 2(Suppl 2): S25–28. doi: 10.1038/ijosup.2012.18

Richardson I. L., Frese S. A. Non-Nutritive Sweeteners and Their Impacts on the Gut Microbiome and Host Physiology. *Frontiers in Nutrition.* 2022; 25(9): 988144. doi: 10.3389/fnut.2022.988144

Suez J., Cohen Y., Valdés-Mas R., Mor U., Dori-Bachash M., Federici S., Zmora N., Leshem A., Heinemann M., Linevsky R., Zur M., Ben-Zeev Brik R., Bukimer A., Eliyahu-Miller S., Metz A., Fischbein R., Sharov O., Malitsky S., Itkin M., Stettner N., Harmelin A., Shapiro H., Stein-Thoeringer C. K., Segal E., Elinav E. Personalized Microbiome-driven Effects of Non-nutritive Sweeteners on Human Glucose Tolerance. *Cell.* 2022; 185(18): 3307–3328.e19. doi: 10.1016/j.cell.2022.07.016. Epub 2022 Aug 19

Watson S. *Dopamine: The pathway to pleasure.* Harvard Health Publishing. 20 July 2021. www.health.harvard.edu/mind-and-mood/dopamine-the-pathway-to-pleasure (accessed September 2022)

Willett A. *Drinkology: The Science of What We Drink and What It Does to Us, from Milks to Martinis.* (Robinson) 2019

Zinöcker M. K., Lindseth I. A. The Western Diet-Microbiome-Host Interaction and Its Role in Metabolic Disease. *Nutrients.* 2018; 10(3): 365. doi: 10.3390/nu10030365

habit 5: tune in to your senses

Antonelli M., Donelli D., Barbieri G., Valussi M., Maggini V., Firenzuoli F. Forest Volatile Organic Compounds and Their Effects on Human Health: A State-of-the-Art Review. *International Journal of Environmental Research and Public Health.* 2020; 17(18): 6506. Doi: 10.3390/ijerph17186506

Banfield-Nwachi M. *Queen of clean Marie Kondo says she has 'kind of given up' on tidying at home.* Guardian online. 30 January 2023. www.theguardian.com/lifeandstyle/2023/jan/30/queen-of-clean-marie-kondo-says-she-has-kind-of-given-up-on-tidying-at-home (accessed January 2023)

Bedrosian T. A., Nelson R. J. Timing of Light Exposure Affects Mood and Brain Circuits. *Translational Psychiatry.* 2017 Jan 31; 7(1): e1017. doi: 10.1038/tp.2016.262

Booth S. *This is your brain on binaural beats.* Healthline. 14 May 2019. www.healthline.com/health-news/your-brain-on-binaural-beats (accessed January 2023)

Buijze G. A., Sierevelt I. N., van der Heijden B. C., Dijkgraaf M. G., Frings-Dresen M. H. The Effect of Cold Showering on Health and Work: A Randomized Controlled Trial. *PLOS One*. 2016 Sep 15; 11(9): e0161749. doi: 10.1371/journal.pone.0161749. Erratum in: *PLOS One*. 2018 Aug 2;13(8): e0201978

calm tech. *Principles of calm technology*. calmtech.com (accessed July 2022)

Case A. *Calm Technology: Principles and Patterns for Non-Intrusive Design.* (O'Reilly) 2015

Coplan R. J., Hipson W. E., Archbell K. A., Ooi L. L., Baldwin D., Bowker J. C. Seeking More Solitude: Conceptualization, Assessment, and Implications of Aloneliness. *Personality and Individual Differences*. 2019; 148: 17–26. doi: 10.1016/j.paid.2019.05.020

DavidHugh. davidhugh.com/en/ (accessed November 2022)

Davis N. *Scientists find part of brain responds selectively to sound of singing.* Guardian online. 22 February 2022. www.theguardian.com/science/2022/feb/22/scientists-find-part-of-brain-responds-selectively-to-sound-of-singing (Accessed July 2022)

Dingle G. A., Sharman L. S., Bauer Z., Beckman E., Broughton M., Bunzli E., Davidson R., Draper G., Fairley S., Farrell C., Flynn L. M., Gomersall S., Hong M., Larwood J., Lee C., Lee J., Nitschinsk L., Peluso N., Reedman S. E., Vidas D., Walter Z. C., Wright O. R. L. How Do Music Activities Affect Health and Well-Being? A Scoping Review of Studies Examining Psychosocial Mechanisms. *Frontiers in Psychology*. 2021; 12: 713818. doi: 10.3389/fpsyg.2021.713818

DiSalvo D. *The big stink about anxiety: It changes how our brains process odors.* Forbes. 28 September 2013. www.forbes.com/sites/daviddisalvo/2013/09/28/the-big-stink-about-anxiety-it-changes-how-our-brains-process-odors/ (accessed May 2022)

Embong N. H., Soh Y. C., Ming L. C., Wong T. W. Revisiting reflexology: Concept, Evidence, Current Practice, and Practitioner Training. *Journal of Traditional and Complementary Medicine*. 2015; 5(4): 197–206. doi: 10.1016/j.jtcme.2015.08.008

Ferreri L., Singer N., McPhee M., Ripollés P., Zatorre R. J., Mas-Herrero E. Engagement in Music-Related Activities During the COVID-19 Pandemic as a Mirror of Individual Differences in Musical Reward and Coping Strategies. *Frontiers in Psychology*. 2021; 12: 673772. doi: 10.3389/fpsyg.2021.673772

Field T. Massage Therapy Research Review. *Complementary Therapies in Clinical Practice*. 2016; 24: 19–31. doi: 10.1016/j.ctcp.2016.04.005. Epub 2016 Apr 23

Finlay K. A., Wilson J. A., Gaston P., Al-Dujaili E. A. S., Power I. Post-operative Pain Management Through Audio-analgesia: Investigating Musical Constructs. *Psychology of Music*. 2016; 44(3): 493–513. doi: 10.1177/0305735615577247

Goldsby T. L., Goldsby M. E., McWalters M., Mills P. J. Effects of Singing Bowl Sound Meditation on Mood, Tension, and Well-being: An Observational Study. *Journal of Evidence-Based Complementary & Alternative Medicine*. 2017; 22(3): 401–406. doi: 10.1177/2156587216668109

Hahad O., Prochaska J. H., Daiber A., Münzel T. Environmental Noise-Induced Effects on Stress Hormones, Oxidative Stress, and Vascular Dysfunction: Key Factors in the Relationship between Cerebrocardiovascular and Psychological Disorders. *Oxidative Medicine and Cellular Longevity*. 2019; doi: 10.1155/2019/4623109

Handlin L., Hydbring-Sandberg E., Nilsson A., Ejdeback M., Jansson A., Uvnäs-Moberg K. Short-Term Interaction between Dogs and Their Owners: Effects on Oxytocin, Cortisol, Insulin and Heart Rate – An Exploratory Study. *Anthrozoös*. 2011; 24(3): 301–315. doi: 10.2752/175303711X13045914865385

Hawkins R. D., Hawkins E. L., Tip L. "I Can't Give Up When I Have Them to Care for": People's Experiences of Pets and Their Mental Health. *Anthrozoös*. 2021; 34(4): 543–562. doi: 10.1080/08927936.2021.1914434

Ito E., Shima R., Yoshioka T. A Novel Role of Oxytocin: Oxytocin-induced Well-being in Humans. *Biophysics and Physicobiology*. 2019 Aug 24; 16: 132–139. doi: 10.2142/biophysico.16.0_132

Jarry J. *Reflexology research doesn't put its best foot forward*. McGill University Office for Science and Society website. 7 May 2021. www.mcgill.ca/oss/article/health-pseudoscience/reflexology-research-doesnt-put-its-best-foot-forward (accessed June 2022)

Kagge E. *Silence: In the Age of Noise*. (Viking) 2017

Kempton B. *Wabi Sabi. Japanese Wisdom for a Perfectly Imperfect Life*. (Piatkus) 2018

Kim R. Burden of Disease from Environmental Noise. In: *WHO International Workshop on Combined Environmental Exposure: Noise, Air Pollutants and Chemicals*. Ispra 2007

Kondo M. *The Life-Changing Magic of Tidying Up*. (Vermilion) 2014

KonMari. konmari.com (accessed September 2022)

Krusemark E. A., Novak L. R., Gitelman D. R., Li W. When the Sense of Smell Meets Emotion: Anxiety-state-dependent Olfactory Processing and Neural Circuitry Adaptation. *Journal of Neuroscience*. 2013; 33(39); 15324–15332. Doi: 10.1523/jneurosci.1835-13.2013

Launay J., Pearce E. *Choir singing improves health, happiness – and is the perfect icebreaker*. The Conversation. theconversation.com/choir-singing-improves-health-happiness-and-is-the-perfect-icebreaker-47619 (accessed November 2022)

Li Q. *Forest Bathing. How Trees Can Help You Find Health and Happiness*. (Penguin Life) 2018

McCullough J. E. M., Liddle S. D., Sinclair M., Close C., Hughes C. M. The Physiological and Biochemical Outcomes Associated with a Reflexology Treatment: A Systematic Review. *Evidence-Based Complementary and Alternative Medicine*. 2014; 502123. doi: 10.1155/2014/502123

Pandya A., Lodha P. Social Connectedness, Excessive Screen Time During COVID-19 and Mental Health: A Review of Current Evidence. *Frontiers in Human Dynamics*. 2021; 3: 684137. doi: 10.3389/fhumd.2021.684137

Pesek A., Bratina T. Gong and its Therapeutic Meaning. *Musicological Annual*. 2016; 52(2): 137–161. doi: 10.4312/mz.52.2.137-161

Pfeifer E., Wittmann M. Waiting, Thinking, and Feeling: Variations in the Perception of Time During Silence. *Frontiers in Psychology*. 2020; 11: 602. doi: 10.3389/fpsyg.2020.00602

Rapaport M. H., Schettler P. J., Larson E. R., Carroll D., Sharenko M., Nettles J., Kinkead B. Massage Therapy for Psychiatric Disorders. *Focus (American Psychiatric Publishing)*. 2018; 16(1): 24–31. doi: 10.1176/appi.focus.20170043. Epub 2018 Jan 24

Rasmussen B., Ekholm O. Is Noise Annoyance from Neighbours in Multi-storey Housing Associated with Fatigue and Sleeping Problems? In: *Proceedings of the ICA 2019 and EAA Euroregio: 23rd International Congress on Acoustics, integrating 4th EAA Euroregio.* 2019; 5071–5078

Stiglic N., Viner R. M. Effects of Screentime on the Health and Well-being of Children and Adolescents: A Systematic Review of Reviews. *BMJ Open.* 2019; 9: e023191. doi: 10.1136/bmjopen-2018-023191

Taniguchi K., Takano M., Tobari Y., Hayano M., Nakajima S., Mimura M., Tsubota K., Noda Y. Influence of External Natural Environment Including Sunshine Exposure on Public Mental Health: A Systematic Review. *Psychiatry International.* 2022; 3(1): 91–113. doi: 10.3390/psychiatryint3010008

Weiser M. *The World is Not a Desktop.* Perspectives article for ACM Interactions. 7 November 1993. https://calmtech.com/papers/the-world-is-not-a-desktop.html (accessed July 2022)

habit 6: find space to recharge your ideas

Barton J., Rogerson M. The Importance of Greenspace for Mental Health. *British Journal of Psychiatry International.* 2017; 14(4): 79–81. doi: 10.1192/s2056474000002051

de Bloom J., Ritter S., Kühnel J., Reinders J., Geurts S. Vacation from Work: A 'Ticket to Creativity'?: The Effects of Recreational Travel on Cognitive Flexibility and Originality. *Tourism Management.* 2014; 44: 164–171. doi: 10.1016/j.tourman.2014.03.013

Britton E., Kindermann G., Domegan C., Carlin C. Blue Care: A Systematic Review of Blue Space Interventions for Health and Wellbeing. *Health Promotion International.* 2020; 35(1): 50–69. doi: 10.1093/heapro/day103

Brown S. *The Social Benefits of Blue Space: A Systematic Review.* Environment Agency. October 2020.

Capaldi C. A., Dopko R. L., Zelenski J. M. The Relationship Between Nature Connectedness and Happiness: a Meta-Analysis. *Frontiers in Psychology.* 2014; 5: 976. doi: 10.3389/fpsyg.2014.00976

Grassini S. A Systematic Review and Meta-Analysis of Nature Walk as an Intervention for Anxiety and Depression. *Journal of Clinical Medicine.* 2022; 11(6): 1731. Doi: 10.3390/jcm11061731

Kardan O., Gozdyra P., Misic B., Moola F., Palmer L. J., Paus T., Berman M. G. Neighborhood Greenspace and Health in a Large Urban Center. *Scientific Reports.* 2015; 5: 11610. doi: 10.1038/srep11610

Koselka E. P. D, Weidner L. C., Minasov A., Berman M. G., Leonard W. R., Santoso M. V., de Brito J. N., Pope Z. C., Pereira M. A., Horton T. H. Walking Green: Developing an Evidence Base for Nature Prescriptions. *International Journal of Environmental Research and Public Health.* 2019; 16(22): 4338. doi: 10.3390/ijerph16224338

McCabe C. *The science behind why hobbies can improve our mental health.* 15 February 2021. University of Reading. research.reading.ac.uk/research-blog/the-science-behind-why-hobbies-can-improve-our-mental-health/ (accessed November 2022)

McDougall C. W., Hanley N., Quilliam R. S., Oliver D. M. Blue Space Exposure, Health and Well-being: Does Freshwater Type Matter? *Landscape and Urban Planning.* 2022; 224: 104446

Nippon.com. *Japan's 72 microseasons.* 16 October 2015. www.nippon.com/en/features/h00124/ (accessed May 2022)

Pasanen T. P., White M. P., Wheeler B. W., Garrett J. K., Elliott L. R. Neighbourhood Blue Space, Health and Wellbeing: The Mediating Role of Different Types of Physical Activity. *Environment International.* 2019; 131: 105016. doi: 10.1016/j.envint.2019.105016

RSPB. *Nature Prescriptions: supporting the health of people and nature.* rspb.org.uk/natureprescriptions (accessed August 2022)

Soga M., Gaston K. J., Yamaura Y. Gardening is Beneficial for Health: A Meta-analysis. *Preventative Medicine Reports.* 2017; 5: 92–99. doi: 10.1016/j.pmedr.2016.11.007

Stier-Jarmer M., Throner V., Kirschneck M., Immich G., Frisch D., Schuh A. The Psychological and Physical Effects of Forests on Human Health: A Systematic Review of Systematic Reviews and Meta- Analyses. *International Journal of Environmental Research and Public Health.* 2021; 18(4): 1770. doi: 10.3390/ijerph18041770

Stock S., Bu F., Fancourt D., Wan Mak H. Longitudinal Associations Between Going Outdoors and Mental Health and Wellbeing During a COVID-19 Lockdown in the UK. *Scientific Reports.* 2022; 12: 10580. doi: 10.1038/s41598-022-15004-0

Syrek C. J., de Bloom J., Lehr D. Well Recovered and More Creative? A Longitudinal Study on the Relationship Between Vacation and Creativity. *Frontiers in Psychology.* 2021; 12: 784844. doi: 10.3389/fpsyg.2021.784844

White M. P., Alcock I., Grellier J., Wheeler B. W., Hartig T., Warber S. L, Bone A., Depledge M. H., Fleming L. E. Spending At Least 120 Minutes a Week in Nature is Associated with Good Health and Wellbeing. *Scientific Reports.* 2019; 9(1): 7730. doi: 10.1038/s41598-019-44097-3

Williamson E. *Finally, the real answer why your best ideas come while showering.* UVAToday. 5 October 2022. news.virginia.edu/content/finally-real-answer-why-your-best-ideas-come-while-showering (accessed November 2022)

Wohlleben P. *The Hidden Life of Trees: What They Feel, How They Communicate – Discoveries from a Secret World.* (Greystone Kids) 2016

habit 7: cherish positive relationships

Age UK. *Social connections and the brain.* 20 September 2022. www.ageuk.org.uk/information-advice/health-wellbeing/mind-body/staying-sharp/looking-after-your-thinking-skills/social-connections-and-the-brain/ (accessed October 2022)

BBC Four. *Mission: Joy – with Archbishop Desmond Tutu and the Dalai Lama.* (viewed May 2022)

Bhattacharya K., Ghosh A., Monsivais D., Dunbar R., Kaski K. Absence Makes the Heart Grow Fonder: Social Compensation When Failure to Interact Risks Weakening a Relationship. *EPJ Data Science.* 2017; 6: 1. doi: 10.1140/epjds/s13688-016-0097-x

Buecker S., Mund M., Chwastek S., Sostmann M., Luhmann M. Is loneliness in emerging adults increasing over time? A preregistered cross-temporal meta-analysis and systematic review. *Psychological Bulletin*. 2021; 147(8): 787–805. doi.: 10.1037/bul0000332

Kuss D. J., Griffiths M. D. Online Social Networking and Addiction – A Review of the Psychological Literature. *International Journal of Environmental Research and Public Health*. 2011; 8(9): 3528–3552. doi: 10.3390/ijerph8093528

Lieberman M. D. *Social: Why Our Brains Are Wired to Connect*. (OUP Oxford) 2013

Marshall L. *We're hard-wired for longing, new study suggests*. University of Colorado Boulder. 11 May 2020. www.colorado.edu/today/2020/05/11/were-hard-wired-longing-new-study-suggests (accessed December 2022)

Umberson D., Montez J. K. Social Relationships and Health: A Flashpoint for Health Policy. *Journal of Health and Social Behavior*. 2010; 51(Suppl): S54–66. doi: 10.1177/0022146510383501

habit 8: nurture a sense of purpose and belonging

Alden L. E., Trew J. L. If It Makes You Happy: Engaging in Kind Acts Increases Positive Affect in Socially Anxious Individuals. *Emotion*. 2013; 13(1): 64–75. doi: 10.1037/a0027761. Epub 2012 May 28

Allen K. A., Kern M. L., Rozek C. S., McInerney D., Slavich G. M. Belonging: A Review of Conceptual Issues, an Integrative Framework, and Directions for Future Research. *Australian Journal of Psychology*. 2021; 73(1): 87–102. doi: 10.1080/00049530.2021.1883409. Epub 2021 Apr 30

AndrewNewberg.com. *How do meditation and prayer change our brains?* www.andrew-newberg.com/research (accessed June 2022)

BBC Four. *Mission: Joy – with Archbishop Desmond Tutu and the Dalai Lama*. (viewed May 2022)

Boggiss A. L., Consedine N. S., Brenton-Peters J. M., Hofman P. L., Serlachius A. S. A Systematic Review of Gratitude Interventions: Effects on Physical Health and Health Behaviors. *Journal of Psychosomatic Research*. 2020; 135: 110165. doi: 10.1016/j.jpsychores.2020.110165

Cregg D. R., Cheavens J. S. Healing Through Helping: An Experimental Investigation of Kindness, Social Activities, and Reappraisal as Well-being Interventions. *The Journal of Positive Psychology*. 2022; doi: 10.1080/17439760.2022.2154695

Dalai Lama, Cutler H. C. *The Art of Happiness. A Handbook for Living*. (Hodder Paperbacks) 1999

Dinic M. *The YouGov Death Study: The meaning of life*. YouGov. 6 October 2021. yougov.co.uk/topics/society/articles-reports/2021/10/06/yougov-death-study-meaning-life (accessed December 2022)

Gao J., Leung H. K., Wu B. W. Y., Skouras S., Hung Sik H. The Neurophysiological Correlates of Religious Chanting. *Scientific Reports*. 2019; 9: 4262. doi: 10.1038/s41598-019-40200-w

Heintzelman S. J., Mohideen F., Oishi S., King L. A. Lay Beliefs About Meaning in Life: Examinations Across Targets, Time, and Countries. *Journal of Research in Personality*. 2020; 88: 104003. doi: 10.1016/j.jrp.2020.104003

Nelson-Coffey S. K., Fritz M. M., Lyubomirsky S., Cole S. W. Kindness in the Blood: A Randomized Controlled Trial of the Gene Regulatory Impact of Prosocial Behavior. *Psychoneuroendocrinology*. 2017; 81: 8–13. doi: 10.1016/j.psyneuen.2017.03.025

Otake K., Shimai S., Tanaka-Matsumi J., Otsui K., Fredrickson B. L. Happy People Become Happier through Kindness: A Counting Kindnesses Intervention. *Journal of Happiness Studies*. 2006; 7(3): 361–375. doi: 10.1007/s10902-005-3650-z

Scott S. K., Lavan N., Chen S., McGettigan C. The Social Life of Laughter. *Trends in Cognitive Sciences*. 2014; 18(12): 618–620. doi: 10.1016/j.tics.2014.09.002. Epub 2014 Oct 22

Simão T. P., Caldeira S., Campos de Carvalho E. The Effect of Prayer on Patients' Health: Systematic Literature Review. *Religions*. 2016; 7: 11. doi: 10.3390/rel7010011

Statista. *Daily time spent on social networking by internet users worldwide from 2012 to 2022*. January 2022. www.statista.com/statistics/433871/daily-social-media-usage-worldwide/ (accessed January 2023)

embracing *calmism* in your daily life

Clear J. *Atomic Habits*. (Random House) 2018

Harvey S. *Kaizen: The Japanese Method for Transforming Habits*. (Pan Macmillan) 2019

Imai M. *Gemba Kaizen. A Commonsense Approach to a Continuous Improvement Strategy*. (McGraw-Hill Education) 2012

World Executives Digest. *Why Kaizen allows success for these companies*. 2 November 2020. www.worldexecutivesdigest.com/why-kaizen-allows-success-for-these-companies/ (accessed May 2022)

index